TRADE AND THE ENVIRONMENT

Trade and The Environment

Analysis of Reciprocal Interactions

by

Savaş Alpay
Beykent University,
Department of International Trade,
Istanbul, Turkey

SPRINGER–SCIENCE+BUSINESS MEDIA, B.V.

A C.I.P. Catalogue record for this book is available from the Library of Congress.

ISBN 978-1-4020-7347-2 ISBN 978-1-4615-0271-5 (eBook)
DOI 10.1007/978-1-4615-0271-5

Printed on acid-free paper

Contents

List of Tables

Acknowledgments

I would like to thank Tarık Kara, Salih Ünlü, İbrahim Yılmaz, Harun Çelik, İlker Dastan, İbrahim Bakır, and people at Kluwer. Tarık's help on technical matters regarding Latex was immensely useful. Salih provided excellent research assistance. İbrahim, Harun, İlker and İbrahim helped me with proofreading. The Latex help desk at Kluwer was prompt and very helpful. It was a great pleasure to work with Cathelijne van Herwaarden at Kluwer. Last but not least, I can not overemphasize the support of my family. Without their support, this work would not be possible.

Chapter 1

INTRODUCTION

With the recent emergence of global environmental issues such as ozone-layer depletion, the loss of biodiversity, global warming and acid rain, the role of economics both in the causes of these problems and in the development of policy options are being examined thoroughly. The new environmental phenomenon with its global scale poses significant and serious threats on future generations. A particular branch of the studies on the economics of the environment analyzes the two-way interaction between trade and the environment. It may seem that the work on trade and environment is very recent as a result of the increased public awareness on actual and potential threats to global environment; nevertheless, the pioneers of this literature published their work as early as the beginning of 1970s (see, for example, Baumol 1971, Magee and Ford 1972, and Walter 1973). This literature is now quite substantial. We will explore different aspects of the trade and the environment literature in this monograph. It will be useful to see some of the trends regarding global environmental problems first.

1. Selected Environmental Trends

Panayotou (1993a) gives a concise definition of environment and environmental degradation: "The term 'environment' refers to both the quantity and quality of natural resources, renewable and nonrenewable. It also includes the ambient environment, which consists of the landscape, water, air, and the atmosphere and constitutes an essential element of the quality of life. Defined as such, the environment is a critical determinant of the quantity, quality, and sustainability of human activities and life in general. Environmental degradation then is the diminution of the environment in quantity, and its deterioration

S. Alpay (ed.), Trade and The Environment, 1–9.

in quality[1]."

The main reason for environmental degradation is the lack of clearly defined *property rights* for environmental resources. As common resources, they are subject to overexploitation: seemingly "innocent" uses of these resources by individuals, when aggregated, jeopardize their own well-being, in the line of "*tragedy of commons*" associated with social ecologist Garrett Hardin. Let us list some facts that reflect the relevance of tragedy of commons for global environment today.

Since the first Earth Day in 1970, nearly 200 million hectares of tree cover were lost; 120 million hectares of land turned into desert area. An estimated 480 billion tons of top soil were destroyed, and many different living species are not with us anymore[2]. Global fishing harvest declined 6%, from 83 million to 77 million tons, in between 1989 and 1991, as a result of nonsustainable fishing practices around the world[3].

Another major problem has been the depletion of the ozone layer. The thickness of the protective ozone layer over the middle latitudes has declined by 3 to 5 percent since 1980 (the decline over Antarctica was 50 percent). An additional 2 percent decline is expected before the attempts through Montreal protocol to reverse this trend become effective. The US Environmental Protection Agency estimated that 350 million cases of skin cancer and 7.5 million fatalities would be expected, due to high levels of ultraviolet radiation, by the year 2075, without the Montreal Protocol to reduce ozone-depleting CFC use[4].

Some 70 million tons of sulfur and 27 million tons of nitrogen in the form of nitrogen oxides were released into the atmosphere in 1991. This corresponds to almost 300% increase in nitrogen and 130% increase in sulfur since 1950[5]. Carbon dioxide level has risen by 26% in the last century. Some scientists project that due to a doubling of greenhouse gases (especially of carbon dioxide) in the atmosphere by the year 2050, average global temperature will reach to critical levels, and this increase potentially brings about higher sea levels, loss of coastal wetlands, severe storms and floods, conversion of agricultural lands into deserts, and threatens the survival of many animal and plant species[6]. The main contributor to global warming is carbon dioxide (50 percent of total

---

[1]Panayotou (1993a), chapter 1, p. 3.
[2]See Brown *et al.* (1991).
[3]UNFAO 1993.
[4]Environmental Protection Agency (EPA) 1993, Global Change Division.
[5]Brown *et al.* 1994.
[6]IPCC 1990, xxii-xxiii.

Table 1.1. Contributions to Global Warming: Across Activities and Greenhouse Gases

	Greenhouse Gases					
Activity	Carbon Dioxide	Methane	Ozone	Nitrous Oxide	CFC	Percent Warming by Activity
Energy Use	35	4	6	4	–	49
Deforestation	10	4	–	–	–	14
Agriculture	3	8	–	2	–	13
Industry	2	-	2	–	20	24
Percent Warming by Gas	50	16	8	6	20	100

Source: World Resources 1990–91, World Resources Institute, as reprinted in Arnold and Long (1994).

warming), followed by CFCs (20 percent), methane (16 percent), ozone, and nitrous oxide. Energy usage makes up 70 percent of total carbon dioxide release, and almost 50 percent of global warming. Other main activities that generate greenhouse gases, and thus, cause global warming, are industrial production, deforestation, and agriculture. Details are presented in table 1.1.

Deforestation in tropics, which hold almost 60 – 70% of the world's biodiversity and help prevent climatic change and protect water supplies, has become a major problem. Ninety percent of the deforestation that is seen in tropics, is for the clearing of land to grow cash crops, most of which is marketed *internationally*. The Amazon forest is being cleared in Ecuador for the extraction of petroleum by US companies, which is again taken to international markets. In Brazil, source of wood, such as mahogany of which 50% is exported to UK, is the Amazon forest. The same place is also cleared to grow coffee and soy beans mostly for the international market[7].

South-eastern Asian countries, the proponents of the so-called outward orientation policies are now facing severe environmental problems: South Korea became one of the highest consumers of pesticides in the world on a per hectare basis. Its capital, Seoul, is "consistently" keeping its place in top five among the world's major cities with threatening levels of sulfur dioxide, with 87% of emissions coming from pollution-intensive industries, rather than from car pollution. Taiwan is not much different. The rate of industrial density in Taiwan, three factories per square kilometer, is 75 times higher than that of the USA. The country has the world's highest rate of hepatitis infection; asthma

[7]These statistics are from Chichilnisky (1994a).

and cancer cases are increasing very rapidly.[8]

Finally, we would like to cite following indicators as reported in the "World Resources 1998-99: A guide to the Global Environment", prepared by the World Resource Institute, the United Nations Environmental Program, the United Nations Development Program, and the World Bank collaboratively (reprinted in Nordstrom and Vaughan 1999):

> Global energy use has increased nearly 70 percent since 1971, and is projected to increase at more than 2 percent annually over the next 15 years. This will raise greenhouse gas emissions by 50 percent over current levels unless a concerted effort is made to increase energy efficiency and move away from today's heavy reliance on fossil fuel.
>
> The consumption of ozone-depleting substances have gone down by 70 percent since the signing of the Montreal Protocol in 1987. Yet, it will still take another 50 years before the ozone layer has returned to normal levels provided that all countries live up to their commitments. A significant black market and trade in CFCs and other ozone-depletion substances is endangering some of the progress already made.
>
> While acid rain is on the decline in many developed countries due to more stringent regulations on sulphur dioxide and nitrogen oxide emissions, the trend is on the rise in many developing countries. In Asia, sulphur dioxide emissions will double by 2020 if current trends continue.
>
> In the past 50 years, excess nitrogen – principally from fertilizers, human sewage and the burning of fossil fuel – has begun to overwhelm the global nitrogen cycle, with a variety of ill effects ranging from reduced soil fertility and overfeeding of lakes, rivers and costal waters. At the current trend, the amount of biologically available nitrogen will double in 25 years.
>
> Deforestation shows no sign of abating. Between 1960 and 1990, some 20 percent of all tropical forests in the world were cleared. In the Amazons alone, some 20000 square kilometers are cleared every year. A leading cause of deforestation in developing countries is extension of subsistence farming and government-backed conversion of forests to large scale ranching and plantations. At the same time, the forest cover in developed countries is stable or even increasing slightly. However, natural forests (that have never been logged) still lack adequate protection in many places.
>
> Bio-diversity is threatened in many places, not just because of a reduction in the habitats as forests are cleared but also because of pollution. Another reason is the competition from non-native plants introduced by humans. Some statistics suggest that 20 percent of all endangered species are threatened by so-called "exotic invaders".
>
> The aquatic environment and its productivity are on the decline. Some 58 percent of the world's coral reefs and 34 percent of all fish species are currently at risk from human

[8]See Khan (1996).

activities. Most oceans are already over-fished with declining yields.

Global water consumption is rising quickly, and the availability of water is likely to become one of the most pressing issues of the 21st century. One third of the world's population lives in areas already experiencing moderate to high water shortages, and that number could (at given population forecasts) rise to two thirds in the next thirty years without serious water conservation measures.

2. Interactions Between Trade and the Environment

The interest in trade and the environment goes beyond academics. The international community started to address these issues at the Stockholm Conference in 1972. In the 1980s, the world witnessed a very successful international environmental agreement: The Montreal Protocol on the Substances that Deplete the Ozone Layer. This protocol with its trade provisions as an enforcement tool presents a prominent example for the solution of other global problems such as climate change. The Uruguay Round of the GATT included heated debates between developing and developed countries on environmental issues; however, there were no significant action plans except for the establishment of Committee on Trade and Environment (CTE) under the World Trade Organization. CTE is accordingly charged with investigating methods for how conflicts among trade liberalization, economic development and environmental protection can be resolved. More recently, United Nations (UN) organized a conference on climate change in Kyoto in December 1997. As opposed to earlier UN conferences on environment, some concrete steps were taken in this conference. The participants agreed to sign the Kyoto Protocol, which, for the first time, set legally binding numerical targets aiming at limitations and reductions of greenhouse gases emissions for the developed countries. Real gains are yet to be seen especially after the recent opposition by US.

Earlier, Dean (1992), Beghin et al. (1994), and Ulph (1994) surveyed the literature on trade and the environment. Ekins et al. (1994) edited a collection of papers on the same subject. Dean's (1992) survey looks at many different aspects of trade and the environment literature: (i) international competitiveness and environmental regulation with subheadings such as relocation of industries to pollution havens, regulation and comparative advantage; (ii) transnational pollution and trade; (iii) product standards as non-tariff barriers; (iv) trade in hazardous substances; and (v) the implications of trade liberalization for environmental degradation and natural resource use. Beghin et al. (1994) update Dean's study on items (i) and (v), and they also introduce the literature on the linkages between trade and environmental policy instruments. Ekins et al. (1994) present a critical assessment of the gains from trade argument from, mainly, an ecological perspective, and offer suggestions for ecologically-

accelerated trade deregulation.

A more-specialized survey of this literature in regards to international competitiveness versus environmental regulation is presented by Jaffe et al. (1995). There is a considerable amount of research in this area, but not a solid consensus on the impact of environmental regulations on the international competitiveness. Jaffe et al. (1995) conclude "International differences in environmental regulatory stringency pose insufficient threats to U.S. industrial competitiveness to justify substantial cutbacks in domestic environmental regulations. Nor does the evidence recommend enactment of stricter domestic environmental regulations in order to stimulate economic competitiveness".

Van Beers and van den Bergh (1996) present an overview of the methodological approaches for investigating the diverse and multi-faceted relationship between international trade and environmental externalities. They classify the studies according to the type of their models (general equilibrium or partial equilibrium models), market structures (perfect or imperfect) and the way environment is treated in the model (as an externality or as an input to the production function). The limitations of each approach are disclosed, and suggestions for improvement are indicated.

Lee (1996) analyzed 79 categorical cases on disputes related to trade and environment. Among these disputes, 24 of them ended in an agreement and 55 in disagreement. Lee concludes that it is not unusual for richer countries to import environmental resources from poorer countries, and that there is a clear indication that issues in developing countries do not receive adequate attention and that richer countries use economic leverage to change the behavior of poor countries.

A final area where trade and environment come together is international environmental agreements. Current trends in the global dimension of the environmental problems necessitate the joint action of countries. In the absence of a supranational authority, solutions to global problems are subject to free-riding, and thus, international environmental agreements are very difficult to achieve and sustain. Related to our subject matter, a particular branch of this literature investigates whether international trade policies may/should be used to enforce international environmental agreements. This line of research, also known as "issue linkage", is analyzed by Folmer et al. (1993), and more recently by Barrett (1997, 1999).

3. Focus of the Monograph

Although trade and environment are interrelated to each other in many different ways, in this monograph, we will explore three main aspects: (i) the impact of trade liberalization on the environment; (ii) the impact of environmental policies on international competitiveness of the firms subject to those policies; and finally, (iii) the interactions between trade, global environmental protection and free-riding especially in regards to international environmental agreements. In each case, we will analyze the issues both theoretically and empirically.

Trade liberalization, through exchange rate policies and removal of tariffs and other trade barriers, will affect environment by its impact on the volume and location of global production and consumption activities. In this regard, some view trade as detrimental to environment since it increases consumption and production, and some others consider it favorably for it will generate more income, and so, demand for environmental quality will rise.

Brown (1990), Ritchie (1990), and Shrybman (1990) argue that trade liberalization inevitably leads to increase in environmental degradation. The statistics on deforestation in Brazil, and Amazon areas, for producing goods for international markets, the deterioration in the environment of outward oriented countries like South Korea, Taiwan (presented above) provide evidences for this standard argument. Supporters of this view suggest that additional trade restrictions are the only way for controlling environmental problems.

On the other hand, Anderson (1992a), using a partial equilibrium analysis, demonstrates that the impact of trade liberalization on a country's environment depends on a number of factors: the country's trade balance; whether the pollution problem is associated with production or consumption; whether trading partners also undertake environmental regulation; and whether or not the country is large enough to influence the price of traded commodities. On the empirical side, Anderson (1992b) shows that liberalizing global trade in coal and food products will reduce global pollution associated with these products.

Trade liberalization may also result in the "importation" of higher environmental standards by developing countries from industrialized countries. An empirical study on Latin American countries by Birdsall and Wheeler (1992) showed that more open-economy countries experienced faster growth in clean industries. One final point of interest is about the substitutability of free trade for some of the traditional environmental policies. Hester et. al. (1993) present supporting evidence. They demonstrate that a move toward freer trade would not only help to meet trade objectives but also would probably provide greater environmental benefits than land set-asides and reduced livestock densities.

In brief, the links between trade liberalization and environment are complex. As will be shown in detail in the first part, both theoretical and empirical studies present mixed evidence. Therefore, it will be a futile attempt to try to search for a universal result regarding the impact of trade liberalization on the environmental quality. One needs to consider many country-specific factors. In the first part of our monograph, we will try to disclose the linkages between trade and environment in this context.

In the second part, we reverse the direction of causality and analyze the impact of environmental policies on international competitiveness and so, on the pattern of international trade.

Conventionally, it has been argued that environmental regulations would harm the competitiveness of the firms being regulated as compared to those subject to lax environmental conditions. Recently, this view has been challenged by a revisionist school. They argue that properly crafted environmental regulations (i.e., incentive-based regulations such as effluent fees, and tradable permits) not only bring social benefits (like increased environmental quality, decline in health risks associated with pollution, etc.), but also can increase competitiveness of the firms being regulated since higher environmental standards can trigger innovation that may offset the compliance costs (this view is also known as the Porter hypothesis).

In chapters 9 to 11, we will focus on this debate between revisionist school and conventional school, and present theoretical and empirical evidences for both. An interesting finding of this part is that pro-revisionist theoretical results do not call for assumptions outside the neoclassical world as suggested by most of the earlier studies in this literature.

In the last part of this monograph, we turn our attention to interactions among global environmental protection, free-riding and international trade. International cooperation on global environmental protection through international environmental agreements (IEA) is a very difficult process to realize, but what is equally difficult is their implementation. In the absence of a supranational authority, IEAs are subject to free-riding, and thus, their chance of being successful is very limited if a proper enforcement mechanism is not developed. An interesting suggestion is the use of trade policies as enforcement tools in the implementation of international environmental agreements.

Firstly, we will analyze (i) the current international trading framework under GATT and its provisions for environmental protection; (ii) international treaties on global environmental problems; (iii) regional trade blocs and their environmental provisions as in the example of NAFTA. Then, we will study global environmental protection in the presence of free-riding in a Ricardian trade model. Since global protection confers benefits on all countries regardless of their contribution (presence of free-riding), we try to determine (i) how-much each country will contribute to global protection, if any; (ii) the factors which affect the level of contributions; (iii) comparison of contributions to global environmental protection when countries do not cooperate versus when they cooperate; and finally, (iv) the impact of free trade on these results.

I

ENVIRONMENTAL CONSEQUENCES OF TRADE LIBERALIZATION

Chapter 2

MODELS OF TRADE AND ENVIRONMENT

One of the standard- for some economists, the deepest and most beautiful- results of international trade is that free trade will benefit all participants. That is why trade liberalization is considered to be a very important step towards development by many economists and politicians. Nevertheless, as indicated by Corden (1974), "Theory does not 'say' that 'free trade is best'. It says that given certain assumptions, it is 'best'." As environmental problems were not a big concern at the time when Samuelson's gains from trade arguments were introduced (see, for example, Samuelson 1962), one such (implicit) assumption was that the environmental impact of trade liberalization policies could be neglected. With the recent emergence of environmental consciousness, the gains from trade argument is being questioned deeply from an environmental perspective by taking the environmental impact of freer trade into consideration. As Dean (1992) puts it, at the center of debate is whether or not the trade reforms will lead to depletion of non-renewable resources and increased environmental degradation, i.e. a type of development which can not be sustained. Thus, it is essential to identify the environmental consequences of trade liberalization. This is a quite demanding task.

A complex web of interactions between trade liberalization and environmental quality makes it difficult to draw clear-cut conclusions. One needs to consider following set of issues: (i) whether as a matter of positive economic analysis one can predict whether a move to trade liberalization, all else constant, will degrade the environment; (ii) how such a prediction might be affected if trade liberalization is accompanied by endogenously determined environmental policies; (iii) whether environmental policies are affected by strategic trade considerations; (iv) whether empirical evidence supports the predictions of eco-

S. Alpay (ed.), Trade and The Environment, 13–16.

nomic analysis; and finally, (v) how all these concerns are affected in case of an environmental problem with international dimension, that is to say, transboundary (or transnational) pollution. In the following chapters, the channels through which further liberalization affects environment will be identified. Two contesting views that trade liberalization will bring improvement versus deterioration in environmental conditions, will be analyzed.

The fundamental determinants of environmental degradation are (i) economic activities (consumption–production) and (ii) the presence, stringency and implementation of national environmental policies. Some environmental problems are consumption-based, and pollutants are generated during or after consumption, such as waste packages, bottles etc. Most other environmental problems are generated during the production process. Obviously, international trade by itself is not a direct source of environmental degradation or improvement. It may be considered as an indirect source due to its potential impact on consumption and production, and on national environmental policies. This potential impact depends on many country-specific factors, and thus can be sometimes positive and sometimes negative. Before moving into the detailed exposition of methods of analysis and findings of different studies, a brief introduction to the various linkages between trade liberalization and the environment will be useful.

Trade liberalization will most likely increase the volume of economic activities (including the dirty ones) and thus, will generate higher environmental degradation. Moreover, the pressures associated with the competition of domestic firms with their foreign rivals may lead to relaxation of the stringent environmental policies or may delay the enactment of stringent environmental policies if they do not exist at the time of trade liberalization.

On the other hand, trade liberalization is also expected to change the composition of goods and services produced by the country; this composition effect may be such that the share of pollution intensive commodities in total output may decline. In addition to this, environmental policies of higher-standard countries may be imported to lower-standard ones either voluntarily with the increases in interactions among countries, or as a result of new regulations on the harmonization of the standards among trading countries. Similarly, the threat to impose trade sanctions may also induce better environmental performance. Furthermore, with higher trade liberalization, if countries can realize expected increases in their incomes, then demand for environmental quality will rise, and this will generate public pressure for improvement in environmental conditions.

The theoretical and empirical methodologies employed in the analysis of environmental consequences of freer trade are varied. As compared to theoretical approaches, empirical studies are more homogeneous in their modelling; two common approaches have been employed in most of the empirical studies: (i) Computable General Equilibrium (CGE) models, and (ii) models with single reduced form equations (missing structural equations in the background) incorporating some common parameters such as pollution, income, and indicators of openness. The latter one is the generic modelling of the environmental Kuznets curve (EKC) studies. Some other papers applied simulation techniques to identify the impact of hypothesized trade liberalization policies on the environment. Proceeding chapters will introduce some of the theoretical and empirical models mostly used in this literature.

The theoretical literature on the environmental consequences of international trade is quite dense, and many interesting sides of the subject have been modelled. It is difficult to single out a common pattern in the theoretical approaches; we see a wide spectrum of models ranging from partial equilibrium ones to general equilibrium ones (which, in themselves, also show big variations, from two-sector to the continuum of good ones). Some papers focus on the supply side of the economy, and model environmental resources as an input into the production processes; others use demand for environmental quality as an additional variable in the consumers' utility function. Modelling differences among the papers in this literature can be grouped under the following headings:

- Partial equilibrium versus general equilibrium
- Imperfect markets versus perfectly competitive markets
- Large country case versus small-country case
- Existence or abscence of environmental regulations
- Governments determine environmental policies strategically versus optimally
- Type of the regulatory instrument used in environmental regulations (standards, tradeable permits or taxes)
- Existence of possibilities for strategic behavior by firms (such as investment in R&D)
- Dynamic versus static set-up
- Transnational (or global) pollution versus local pollution
- Pollution generated during or after consumption versus production

- Incorporation of environmental resources as inputs versus by-products of production
- Global or regional focus in regards to the coverage of countries

This is not an all exhaustive list; however, main stream analysis has been carried out along the dimensions listed above. Differences in conclusions, sometimes contrasting, of the studies in this literature stem from these different modelling assumptions. Although partial equilibrium framework is a subset (or a special case) of general equilibrium modelling, there are very interesting results and insights derived in a partial equilibrium set-up, which may be lost in a general equilibrium approach; thus, partial equilibrium aproaches deserve attention and are worth studying. We will present our theoretical exposition along the partial and general equilibrium separation. Most of the distinctions along the lines listed above will be explored in each case.

Before moving to modelling exposition, a general remark on how the environmental consequences of trade liberalization can be identified theoretically, will be useful. One can talk about two approaches. The obvious one is to compare the level of emissions before the trade liberalization (i.e. in autarky, or less liberalized situation) and after it. The second one is to determine the impact of trade liberalization (or the existence of trade) on the level of environmental policies; that is to say to identify whether regulation agencies relax the stringency of environmental policies (known as "eco-dumping" or "race to the bottom" in the literature) during trade liberalization. This second approach also has connections with the impact of environmental regulations on the competitiveness of the regulated firms, which is the second theme of this monograph, as the relaxation of environmental regulations arises from the incentives to confer competitive advantage on the own firms. Extensive analysis of such interactions will be presented in the second part.

A compact presentation of both the partial and general equilibrium models will be given in the proceeding chapters. We start with the partial equilibrium models.

Chapter 3

PARTIAL EQUILIBRIUM MODELS OF TRADE AND ENVIRONMENT

In this chapter, different types of partial equilibrium models of trade and environment will be presented. In each section, both the welfare and the environmental impact of (further) trade liberalization policies will be identified. The impact of changes in the models such as small versus large country, consumption versus production externality, perfectly versus imperfectly competitive markets, presence versus abscence of environmental policies will be demonstrated.

1. Production Externality in a Small Country without Environmental Policy

Assume that production of a good generates local pollution. The domestic demand and supply functions for this good for a given price level P are given by $D(P)$ and $S(P)$, respectively. For simplicity it is assumed that one unit of output generates one unit of emission of the pollutant, so emissions will be given by $E = S(P)$. The welfare cost of pollution is $C(E)$ or $C(S(P))$. In autarky the welfare level, which is the sum of producer and consumer surplus net of pollution cost is as follows:

$$W_A = \int_P^{\infty} D(x)dx + \int_0^P S(x)dx - C(S(P)) \tag{3.1}$$

where W_A refers to autarky welfare level, and P is the domestic equilibrium price. Suppose that world prices, P_W are below the domestic price level, i.e.

S. Alpay (ed.), Trade and The Environment, 17–29.

$P_W < P$. Then this good will be imported by this country when it gets openned to international markets. The welfare level after trade will be given by:

$$W_T = \int_{P_W}^{\infty} D(x)dx + \int_{0}^{P_W} S(x)dx - C(S(P_W)) \tag{3.2}$$

The change in the welfare and emissions of the country can easily be determined, by using equations (3.1) and (3.2), as:

$$E_T - E_A = S(P_W) - S(P) \tag{3.3}$$

$$\begin{aligned} W_T - W_A &= \int_{P_W}^{P} D(x)dx - \int_{P_W}^{P} S(x)dx + C(S(P)) - C(S(P_W)) \\ &= \int_{P_W}^{P} (D(x) - S(x))dx + C(S(P)) - C(S(P_W)) \end{aligned} \tag{3.4}$$

where E_T represents the level of emissions after trade liberalization. From (3.3), as $S' > 0$, and $P > P_W$, the level of emissions falls after trade as expected since the country is importing the good. Similarly, the welfare level after trade exceeds that of the autarky from (3.4): Since there will be excess demand (at home) for prices below P, (note that $P_W < P$, and P represents the equilibrium price before trade), so the integral expression in (3.4) will be positive, and the welfare cost of emissions will fall as the level of emissions is lower after trade.

The previous analysis for a small country can easily be extended to the case where the domestic price is initially lower than the world price, i.e. $P < P_W$. In this case, the country will export this good. The change in the level of emissions will be same as the one in equation (3.3); however, as $P < P_W$, after trade emissions will increase (due to higher production associated with exporting). The welfare change in equation (3.4) will be slightly modified:

$$\begin{aligned} W_T - W_A &= -\int_{P}^{P_W} D(x)dx + \int_{P}^{P_W} S(x)dx + C(S(P)) - C(S(P_W)) \\ &= \int_{P}^{P_W} (S(x) - D(x))dx + C(S(P)) - C(S(P_W)) \end{aligned} \tag{3.5}$$

This time the welfare level after trade may exceed or fall below that of the autarky. This can easily be seen from equation (3.5). There will be excess supply (at home) for prices above P, (note that $P_W > P$, and P represents the equilibrium price before trade), and thus, the integral term is always positive; however, the cost of emissions will increase as the level of emissions is higher after trade. Therefore the impact of trade on the welfare of an exporting small country is ambiguous. If the welfare cost of higher emissions exceeds the benefit arising from exports than there will be a decline in the welfare of the country after trade liberalization. Hence, an indication of this analysis is that gains-from-trade argument need to be modified in a setting where there are environmental externalities.

In brief, an importing small country will benefit from openness to international markets on two accounts in the presence of production-based environmental degradation: lower emissions and higher consumer surplus (the increase in consumer surplus exceeds the decline in the producer surplus). On the other hand, in a similar situation, an exporting country may gain or lose from trade. Now, we can start changing some components of our basic partial equilibrium set-up. We first consider a switch to consumption-based environmental externality.

2. Consumption Externality in a Small Country without Environmental Policy

In this section, it is assumed that consumption of a good generates local pollution. The domestic demand and supply functions are as in the previous section, and one unit of consumption generates one unit of emission of the pollutant, so emissions will be equal to $E = D(P)$. Then, the welfare cost of pollution becomes $C(D(P))$. In autarky, the welfare level, which is the sum of producer and consumer surplus net of pollution cost, is given by:

$$W_A = \int_P^\infty D(x)dx + \int_0^P S(x)dx - C(D(P)) \tag{3.6}$$

where W_A refers to autarky welfare level, and P is the domestic equilibrium price. We again start with the case of imports and assume that world price, P_W is below the domestic price level, i.e. $P_W < P$. The welfare level after trade liberalization will be given by:

$$W_T = \int_{P_W}^{\infty} D(x)dx + \int_{0}^{P_W} S(x)dx - C(D(P_W)) \quad (3.7)$$

Changes in the emission and welfare levels will be equal to:

$$E_T - E_A = D(P_W) - D(P) \quad (3.8)$$

$$\begin{aligned} W_T - W_A &= \int_{P_W}^{P} D(x)dx - \int_{P_W}^{P} S(x)dx + C(D(P)) - C(D(P_W)) \\ &= \int_{P_W}^{P} (D(x) - S(x))dx + C(D(P)) - C(D(P_W)) \quad (3.9) \end{aligned}$$

From (3.8), as $D' < 0$, and $P > P_W$, the level of emissions increases after trade as expected since the domestic consumption rises with imports and pollution is consumption based. For the comparison of welfare levels, we refer to equation (3.9). There will be excess demand (at home) for prices below P, (note that $P_W < P$, and P represents the equilibrium price before trade), and so the integral expression in (3.9) will be positive ($D(x) > S(x)$). On the other hand, welfare cost of emissions will increase as the level of emissions is higher after trade. Hence, the impact of trade on the welfare of a small country importing a good, the consumption of which generates environmental degradation, is ambiguous. If the welfare cost of higher emissions exceeds the increase in consumer surplus net of producer surplus arising from imports, then there will be a decline in the welfare of the country after trade liberalization.

Let us now consider the case of exports, which requires that the domestic price is initially lower than the world price, i.e. $P < P_W$. The difference between the level of emissions before and after trade will be given by equation (3.8); however, as $P < P_W$, emissions will decrease after trade (due to lower domestic consumption as a result of higher prices in the presence of exports). The impact on the domestic welfare will be as follows:

$$\begin{aligned} W_T - W_A &= -\int_{P}^{P_W} D(x)dx + \int_{P}^{P_W} S(x)dx + C(D(P)) - C(D(P_W)) \\ &= \int_{P}^{P_W} (S(x) - D(x))dx + C(D(P)) - C(D(P_W)) \quad (3.10) \end{aligned}$$

Note that there will be excess supply (at home) for prices above P, ($P_W > P$, and P represents the equilibrium price before trade), and thus, the integral term is always positive. Moreover, the difference between the cost of emissions (the second term in equation 3.10) will also be positive as the level of emissions is lower after trade. Thus, there will be an unambiguous improvement in the welfare after trade liberalization.

Comparing the results of this section with the previous one, we can immediately notice that there is a perfect asymmetry between the production- and consumption-based environmental externality. In case of environmental degradation arising from production, importing small country benefits from trade in two ways: lower emissions and higher welfare, and so the usual gains-from-trade argument follows. However, an exporting small country ends up with higher emissions, and the impact on the welfare is ambiguous. On the other hand, if the negative environmental externality is consumption-based, then the exporting country benefits from trade, and the importing small country may observe an increase or a decrease in its welfare depending on the cost of higher emissions. Therefore, this simple analysis presents the importance of the source of the environmental externality. We continue to change the basic set-up, and now, we consider the case of a large country.

3. Production Externality in a Large Country without Environmental Policy

As opposed to small countries, large countries can affect the world supply and demand curves. When a large country starts importing a commodity in the world market, the price of the commodity will be driven up (demand increases in the world market), and similarly in case of exports, the world price will go down (supply increases in the world market). Therefore, changes in the world prices have to be incorporated into the analysis shown for a small country above. Note that autarky and world prices were assumed to be equal to P, and P_W, respectively, in the small country case. If initially, $P > P_W$, then the country will be importing the commodity from abroad, but as it is a large country, equilibrium will be established at a higher world price, P'_W, which will satisfy $P > P'_W > P_W$. Rest of the analysis will be same as the small country case with production-based environmental externality. The welfare level after trade will be now given by:

$$W_T = \int_{P'_W}^{\infty} D(x)dx + \int_{0}^{P'_W} S(x)dx - C(S(P'_W)) \qquad (3.11)$$

Changes in the welfare and pollution levels can easily be determined, by using equations (3.1) and (3.11), as:

$$E_T - E_A = S(P'_W) - S(P) \tag{3.12}$$

$$\begin{aligned} W_T - W_A &= \int_{P'_W}^{P} D(x)dx - \int_{P'_W}^{P} S(x)dx + C(S(P)) - C(S(P'_W)) \\ &= \int_{P'_W}^{P} (D(x) - S(x))dx + C(S(P)) - C(S(P'_W)) \end{aligned} \tag{3.13}$$

From (3.12), as $S' > 0$, and $P > P'_W$, the level of emissions falls after trade as expected since the country is importing the good; however the decline in pollution level is smaller as compared to the small-country case. Similarly, the welfare level after trade exceeds that of the autarky from (3.13) because $D(x) > S(x)$ for prices below P, and hence both the integral expression and the difference between environmental costs in (3.13) will be positive. The welfare improvement resulting from trade liberalization for the large country (in 3.13) will fall short of that of a small country (in 3.4) as $P'_W > P_W$.

If the large country becomes an exporter in the world market, which will be the case if initially $P < P_W$, then the equilibrium will be established at a new world price, P'_W, which will satisy $P < P'_W < P_W$. The impact on the level of emission and the welfare, will be as follows:

$$E_T - E_A = S(P'_W) - S(P) \tag{3.14}$$

$$\begin{aligned} W_T - W_A &= -\int_{P}^{P'_W} D(x)dx + \int_{P}^{P'_W} S(x)dx + C(S(P)) - C(S(P'_W)) \\ &= \int_{P}^{P'_W} (S(x) - D(x))dx + C(S(P)) - C(S(P'_W)) \end{aligned} \tag{3.15}$$

As pollution is production-based, the emission level will increase with higher production for an exporting large country. Comparing equation (3.14) with the corresponding difference in emissions for a small country, it can easily be noticed that the increase in emissions will be lower in a large country than a small country as $P'_W < P_W$. The welfare impact is ambiguous as the integral term is always positive ($D(x) < S(x)$ for prices between P and P'_W) and the non-integral expression in (3.15) is negative ($C(S(P)) < C(S(P'_W))$ as $P'_W > P$). Welfare comparison between small and large country cases in the case of exports is not immediately possible. If the decline in the producer surplus net of consumer surplus is compensated by the decline in the fall of pollution cost, then the welfare impact of trade liberalization will be higher in case of a large country. This is the opposite of what we see in case of imports. We now consider the case of consumption-based environmental externality for a large country.

4. Consumption Externality in a Large Country without Environmental Policy

As shown above for the small country case, a switch to consumption-based envrionmental degradation reverses the welfare and pollution impact of trade liberalization for exports and imports. In this section we will analyze the impact of switching to consumption-based environmental degradation for a large country. In autarky the welfare level of a large country is as follows:

$$W_A = \int_P^\infty D(x)dx + \int_0^P S(x)dx - C(D(P)) \tag{3.16}$$

We start with the case of imports, and assume that initially world prices, P_W are below the domestic price level, i.e. $P_W < P$. As above, the world prices will increase to P'_W when large country starts importing this good. The welfare level after trade liberalization will be given by:

$$W_T = \int_{P'_W}^\infty D(x)dx + \int_0^{P'_W} S(x)dx - C(D(P'_W)) \tag{3.17}$$

Changes in the emission and welfare levels will be equal to:

$$E_T - E_A = D(P'_W) - D(P) \tag{3.18}$$

$$W_T - W_A = \int_{P'_W}^P D(x)dx - \int_{P'_W}^P S(x)dx + C(D(P)) - C(D(P'_W))$$

$$= \int_{P'_W}^{P} (D(x) - S(x))dx + C(D(P)) - C(D(P'_W)) \quad (3.19)$$

From (3.18), as $D' < 0$, and $P > P'_W$, the level of emissions increases after trade; however, the increase in the emission level will be less than the corresponding increase in emissions for a small country (see equation 3.8) since $P'_W > P_W$. As in the small country case, the difference between the welfare levels in equation (3.19) can be positive or negative. Domestic demand will exceed supply for prices below the autarky equilibrium price, P, (note that $P'_W < P$), and so the integral expression in (3.19) will be positive. On the other hand, $C(D(P)) < C(D(P'_W))$ due to higher emissions after trade. Hence, the impact of trade on the welfare of a large country importing a commodity with negative consumption externality, is ambiguous. If the welfare cost of higher emissions falls short of the increase in consumer surplus net of producer surplus arising from imports, then the welfare of large importing country will rise after trade liberalization. Finally, the difference between the post-trade welfare change for a large and small country will be given by:

$$\Delta(W_S) - \Delta(W_L) = \int_{P_W}^{P'_W} (D(x) - S(x))dx + C(D(P'_W)) - C(D(P_W)) \quad (3.20)$$

where $\Delta(W_S)$ and $\Delta(W_L)$ refer to the change in the welfare levels of a small and large country, respectively, following trade liberalization. The integral term in equation (3.20) is positive (since $D(x) > S(x)$ in the relevant price range) whereas the non-integral difference is negative. Thus, the relative impact of trade on the welfare levels of a small and large country is ambiguous; however, the post-trade level of emissions will be lower in a large country. We now turn to the case of exports.

If we assume that the domestic price is lower than the world price initially, i.e. $P < P_W$, then the domestic country will export the commodity. As above, the new world equilibrium price, P'_W will be in between P and P_W. The change in the emission level will be same as the one in equation (3.18), but now since $P < P'_W$, the emissions will decline. As opposed to the case of imports, there will be no ambiguity in the comparison of the post- and pre-trade welfare levels:

$$W_T - W_A = \int_{P}^{P'_W} (S(x) - D(x))dx + C(D(P)) - C(D(P'_W)) \quad (3.21)$$

For prices above P, (note that $P'_W > P$), supply is always higher than demand, and thus, the integral term is always positive. Moreover, the difference between the cost of emissions (the second term in equation 3.21) will also be positive as the level of emissions is lower after trade. The comparison of the change in the welfare for the small and large country cases can be seen from:

$$\Delta(W_S) - \Delta(W_L) = \int_{P'_W}^{P_W} (S(x) - D(x))dx + C(D(P'_W)) - C(D(P_W)) \qquad (3.22)$$

where $\Delta(W_S)$ and $\Delta(W_L)$ refer to the change in the welfare levels of small and large country, respectively, following trade liberalization. The integral term in equation (3.22) is positive (since $S(x) > D(x)$ in the relevant price range); similarly, the non-integral difference is also positive as $P'_W < P_W$ and $D' < 0$. Thus, $\Delta(W_S)$ exceeds $\Delta(W_L)$, and so the gains from trade will be higher for a small country than a large country. As opposed to the case of imports, final pollution level is also lower in a small country.

Having presented a simple welfare and pollution analysis in a partial equilibrium framework, we now would like to introduce environmental regulations into our model. A more extensive modification is the introduction of imperfect competition and firm-level details, again in a partial equilibrium set-up. Given that extension to imperfect competition can be better analyzed in a somewhat different set-up, we will leave it to the next chapter. In the final sections of this chapter, we now analyze the impact of environmental regulations.

5. Production Externality in the Presence of Environmental Policy for a Small Country

In this section and the next, we will introduce environmental policy into the analysis presented above for a large and small country. We will start with emission standards as the regulatory instrument, then, we will consider the case in which environmental agency implements its environmental policy through taxation, and charges a product tax per unit of output. Environmental agency determines the emission standards or the tax rate optimally by maximizing the social welfare. There are a few changes in this section regarding some of the basic elements of our set-up in the previous section. Autarky will refer to a situation in which import tariffs or export subsidies are present, and trade liberalization will be identified with changes (reductions) in these variables. We will first consider the case of a small country importing a commodity, the production of which generates pollution. Environmental regulation is assumed to be carried out through emission standards. Assuming that T represents import tariffs per unit of imports, and E represents the emission standards, the welfare level is given by:

$$W_A = \int_P^\infty D(x)dx + PS(P) - C^P(S,E) - C(E) + T\,(D(P) - S(P)) \qquad (3.23)$$

where P is the domestic price level (which is equal to world prices P_W plus import tariffs T), $C^P(S, E)$ represents the cost of production including cost of abatement, and the other functional definitions are same as above. It is assumed that import tariff revenues are distributed among consumers. Environmental agency determines the level of allowable emissions by maximizing the social welfare in (3.23). The first order condition results in

$$\frac{\partial C(E)}{\partial E} = -\frac{\partial C^P(S,E)}{\partial E} \qquad (3.24)$$

Note that the derivative of production cost with respect to emission standard is negative; as emissions decline more abatement have to be undertaken, and this will increase the costs. On the other hand, the environmental damage will go down when standards become tighter (i.e. get reduced). Hence both sides of equation (3.24) are positive. Optimal emission standards before the trade liberalization are given by the solution of (3.24). Now, we assume that the import tariff rate T gets reduced towards liberalizing the trade regime in this country. Since equation (3.24) does not depend on the level of tariffs T, it seems that optimal emissions standards will not change; however, with a decline in tariffs, domestic prices will go down (as world prices are constant), and hence domestic production will be reduced. It is generally assumed that $\frac{\partial^2 C^P(S,E)}{\partial S \partial E} < 0$ in the literature. Under this assumption, right-hand-side of equation (3.24) declines, and so optimal emission standards get reduced if $\frac{\partial^2 C(E)}{\partial E^2} > 0$. If the final condition is reversed then the optimal policy will be modified in the upward direction and allowable emission limits will be increased. Thus higher openness to international markets may have different environmental impact depending on the characteristics of the damage function. This result is somewhat different from the small country case with no environmental policy above. Now we would like to present a more detailed exposition of environmental policy implemented through taxation for a large country. This way differences regarding both the type of regulatory instrument and size of the country will be presented as they relate to the environmental impact of trade liberalization.

6. Production Externality in the Presence of Environmental Policy for a Large Country

We consider the case of an importing large country first. The environmental agency charges environmental production tax t per unit of production. Import tariffs T per unit are applied initially, and they will get reduced with the introduction of trade liberalization policies afterwards. Optimal level of t is determined from the maximization of the following social welfare function:

$$W_A = \int_{P_d}^{\infty} D(x)dx + \int_{0}^{P_s} S(x)dx - C(S(P_s)) + tS(P_s) + T(D(P_d) - S(P_s)) \quad (3.25)$$

where P_d is the domestic price level for consumers, P_s is the domestic price level for producers, tax revenues ($tS(P_s)$) and tariff revenues ($T(D(P_d) - S(P_s))$ are distributed among consumers. Krutilla (1991) studies the optimal environmental policy in a setting similar to this one. He studies both the consumption and production-based externalities, and identifies the optimal environmental and trade policy mix. We will make use of this set-up in our partial equilibrium analysis of the impact of trade liberalization on the environment.

As we are studying a large country, world prices will be affected by its actions. Equilibrium requires that $P_s = P_W + T - t$, and $P_d = P_W + T$, where P_W refers to the world price level. Additionally, trade equilibrium condition must also be satisfied:

$$S(P_s) - D(P_d) = D_W(P_W) - S_W(P_W) \quad (3.26)$$
$$= ED_W(P_W) \quad (3.27)$$

where $S_W(P_W)$, $D_W(P_W)$, and $ED_W(P_W)$ represents the world supply, demand and excess demand functions, respectively, at the world prices.

Environmental agency determines the optimal level of the specific tax rate by maximizing the social welfare in (3.25). The first order condition results in

$$-D(P_d)\frac{\partial P_d}{\partial t} + S(P_s)\frac{\partial P_s}{\partial t} - \frac{\partial C}{\partial S}S'(P_s)\frac{\partial P_s}{\partial t} + S(P_s) + tS'(P_s)\frac{\partial P_s}{\partial t}$$
$$+ T\left(D'(P_d)S(P_s)\frac{\partial P_d}{\partial t} - S'(P)S(P_s)\frac{\partial P_s}{\partial t}\right) = 0 \quad (3.28)$$

Using the price equilibrium conditions, we get

$$\frac{dP_s}{dt} = \frac{dP_W}{dt} - 1 \quad (3.29)$$
$$\frac{dP_d}{dt} = \frac{dP_W}{dt} \quad (3.30)$$

Trade equilibrium in equation (3.27) implies

$$\frac{\partial S}{\partial P_s}\frac{dP_s}{dt} - \frac{\partial D}{\partial P_d}\frac{dP_d}{dt} = \frac{\partial ED_W}{\partial P_W}\frac{dP_W}{dt} \quad (3.31)$$

Using equations (3.29), (3.30) and (3.31) in (3.28), the optimal tax rate in Krutilla (1991) can be derived as:

$$t = \frac{\partial C}{\partial S} + \alpha(S(P_s) - D(P_d)) + \alpha T \frac{\partial ED_W}{\partial P_W} \tag{3.32}$$

where $\alpha = \frac{\frac{dP_W}{dt}}{S'(P_s)\frac{dP_s}{dt}}$ and is non-negative. Equation (3.32) shows that the optimal tax rate is equal to standard Pigouvian tax (the first term on the right hand side) plus terms-of-trade effect (second term) and tariff-revenue effect (last term). For an importing country, terms-of-trade effect is negative, and tariff-revenue effect is positive.

Given these findings from Krutilla (1991), we now introduce a trade liberalization policy, and consider a reduction in the tariffs, T. A reduction in tariffs will have impact both on the price levels and the environmental product tax. To be able to derive analytically tractable expressions, we assume that all demand and supply functions are linear in their arguments. Using the price equilibrium conditions, we get

$$\frac{dP_s}{dT} = \frac{dP_W}{dT} + 1 - \frac{dt}{dT} \tag{3.33}$$

$$\frac{dP_d}{dT} = \frac{dP_W}{dT} + 1 \tag{3.34}$$

From trade equilibrium in equation (3.27), we get

$$S'\frac{dP_s}{dT} - D'\frac{dP_d}{dT} = \frac{\partial ED_W}{\partial P}\frac{dP_W}{dT} \tag{3.35}$$

Solving equations (3.33), (3.34) and (3.35) together, one obtains:

$$\frac{dP_W}{dT} = \frac{S' - S'\alpha ED'_W - D'}{ED'_W - S' + S'\alpha ED'_W + D'} \tag{3.36}$$

Thus, the world price level is a negative function of tariff rates; a reduction in tariffs will increase the world price. From equation (3.34), it can easily be determined that the domestic price, P_d is positively related to tariff rates, hence after trade liberalization they will decrease. To see what happens to the pollution levels we need to compute the impact of tariff reduction on the environmental product tax, t. Using equation (3.32), we obtain:

$$\begin{aligned} \frac{dt}{dT} &= \alpha ED'_W \left(\frac{dP_W}{dT} + 1\right) \\ &= \frac{\alpha (ED'_W)^2}{ED'_W - S' + S'\alpha ED'_W + D'} \end{aligned} \tag{3.37}$$

The expression in (3.37) is nonpositive; trade liberalization thus leads to an increase in the environmental tax rate. Then it is straightforward to show that domestic producers' prices are a positive funcion of tariff rates. A decline in tariff rates leads to a decline in producers' prices, therefore domestic production will decline after trade liberalization policies. As the environmental degradation is generated by production, the impact of freer trade will be an improvement in environmental conditions.

After seeing the impact of trade liberalization policies on the environment for alternative partial equilibrium models which are all perfectly competitive, we will now introduce how these results will be modified in imperfectly competitive markets in the next chapter.

Chapter 4

STRATEGIC BEHAVIOR AND ENVIRONMENTAL QUALITY

In this chapter, the issue of strategic behavior in the presence of environmental regulations and international trade is investigated. Our model will be useful in demonstrating the impact of trade on the environment in imperfectly competitive markets in a partial equilibrium framework. A detailed exposition of interactions between trade and the environment in imperfectly competitive markets can be seen in Rauscher (1997) (see chapters 3,4, 6, and 7). In a two-country, one-good, two-producer model as in Ulph (1996), here we will analyze the Nash equilibrium of the game where governments may behave strategically in choosing their environmental policies, and producers may behave strategically in choosing their R&D investments in the presence of international trade.

1. Introduction

In this chapter, we will focus on an important dimension: strategic environmental policy making in the presence of international trade. Environmentalists worry that trade-related goals may generate some distortions in environmental policies; freer trade may lead to laxer environmental standards in order to generate competitive advantage over the trading partners. This kind of strategic behavior, known as "eco-dumping" or "race to the bottom", attracted the attention of researchers, and a number of studies analyzed this subject.

The studies on strategic behavior usually incorporate extended versions of the Brander and Spencer (1985) model. Brander and Spencer Cournot duopoly model is designed to analyze export subsidies only, and environmental components are added later by Conrad (1993), Kennedy (1994), Barrett (1994b), Rauscher (1994) and Ulph (1996). In Brander and Spencer (1985), the govern-

S. Alpay (ed.), Trade and The Environment, 31–49.

ments in each country maximize the domestic surplus, defined as the profit of the domestic firm net of the subsidy whereas in its extended versions of trade-environment studies, the domestic surplus is usually the profit of the domestic firm net of the environmental damage.

Kennedy (1994) tries to find the optimum pollution tax in open economies. He argues that imperfect competition in global markets creates inefficient distortion of pollution taxes. He investigates two opposite effects that interact with each other: one to gain competitive advantage over the trading partner, and the other to shift the pollution to the other country. It has been shown that in case of perfect transboundary pollution, the second effect vanishes. If the pollution is partially transboundary, this distorts the pollution taxes that would otherwise be globally efficient.

Barrett (1994b) studies the impact of market type on the environmental-policy-making of governments. He shows that if the domestic industry is a monopoly, the foreign industry is imperfectly competitive, and industrial competition is Cournot, then the domestic government has an incentive to set weak environmental standards; weak standards mean that at the level of chosen emissions, the marginal cost of abatement is less than the marginal damage from pollution. Strategically-optimal emission standards are set weaker than the environmentally-optimal emission standards.

Very recently, Tsai (1999) uses a partial equilibrium strategic trade framework to show that trade liberalization can improve environmental quality. In this theoretical study, a very simple international duopoly model is employed. Both the domestic and foreign firms export an identical product to a third market; the production of the good generates pollution which is regulated by the governments. An export subsidy is also given by the home government only. Firms maximize their profits net of abatement expenditure, and governments maximize social welfare inclusive of environmental damage by choosing the optimal emission standards and in the case of home country, per unit export subsidy. Trade liberalization is reflected by a decline in the export subsidy.

Tsai (1999) considers both partial and complete trade liberalization scenarios. In both cases, liberalization leads to better environmental quality. This result is robust to the type of the competition, i.e., it holds under both Cournot and Bertrand competition among the firms. The novelty of this paper is that it suggests comparing the post-liberalization emission levels with the pre-liberalization emission levels instead of optimal emission level, as done in other theoretical pieces in this literature. The post-liberalization emission level turns out to be higher than optimal emission levels but lower than the pre-liberalization emission level. Thus, from an environmental perspective, liberalization is a preferred policy over protection, but it is not the best policy.

Below, we will present the explicit and extensive welfare analysis of the game among producers and governments in the context described in Ulph (1996). Ulph (1996) extends the strategic behavior literature into a setting where not only governments but also producers may act strategically. In a two-country, two-producer and one good model, governments internalize environmental damages associated with the production by either setting standards or levying taxes. Governments may behave strategically by distorting their environmental policies in order to generate competitive advantage to their firms, which are competing in a third country market. Producers' strategic behavior is associated with their choice of R&D investment. Ulph (1996) shows that if producers act strategically, this always reduces the incentives for governments to relax their environmental policies, and that if governments act strategically, this always increases the producers to overinvest in R&D (i.e. act strategically). The paper also demonstrates that when both producers and governments act strategically, the use of pollution taxes is inferior to the use of emission standards (with respect to the distortion in environmental policies and R&D investment), and finally welfare is lower when both producers and governments act strategically than when only one agent acts strategically.

Ulph (1996) did not present a complete analysis of the Nash equilibrium of the strategic game described above. He only considered the symmetric equilibrium. In this part we will present an extension of Ulph's analysis and examine the equilibrium behavior more generally, including non-symmetrical behaviors. This is very critical as the final outcome regarding environmental policies, R&D investment, emission and welfare levels depend on the equilibrium behavior.

The game in the present context can be divided into two stages. In the first stage, governments maximize the domestic welfare associated with the production of the polluting good (equal to producer's surplus net of welfare cost of pollution) by choosing the level of emissions. Governments may behave efficiently or strategically, in which case they will try to induce competitive advantage to their firms by incorporating the impact of the chosen emission levels on the foreign producer's behavior. Similarly, producers may behave optimally or strategically in the context defined in Ulph (1996). Therefore, there will be 16 different set of behaviors, such as both governments are playing strategically and producers behaving non-strategically etc. In our paper we will try to determine the equilibrium behavior. This will be done for both the sequential and simultaneous move games. Such an analysis has not been reported before. The extended version of the present analysis can be seen in Alpay and Karaman (2002).

2. Model

We will analyze the strategic behavior of governments and producers in the presence of environmental regulations and international trade by making use of the model developed by Ulph (1996). This model assumes a two-country set-up with one producer in each of them. The producers produce a homogeneous good, the production of which generates pollution. The producers are competing in a third-country market. The inverse demand function (or price in the third country) is given by $p = A - x - y$, where p is the price, x is the domestic production, y is the foreign production and A is a positive constant. Then the revenue function for the home country producer is given by $R(x, y) = x(A - x - y)$. Countries are assumed to be symmetric, and the terms corresponding to foreign country can thus be imagined easily.

The production cost of x is $C(x, \phi) = \phi x^2/4$, where ϕ represents the technology parameter (or R&D parameter). The producer chooses this strategic variable ϕ, and the cost associated with this choice is given by $\frac{1}{\phi}$ (one can visualize this process as the choice of a technology level, ϕ, and the cost of R&D required to obtain that technology, $\frac{1}{\phi}$). In the absence of strategic behavior, the producer will minimize the total cost function $\phi\frac{x^2}{4} + \frac{1}{\phi}$ for any given level of output to get the efficient level of ϕ. Cost minimization results in $\phi = \frac{2}{x}$. Substituting this into the cost function gives us the total cost:

$$\text{Total cost} = \frac{2}{x}\frac{x^2}{4} + \frac{1}{2/x} = \frac{x}{2} + \frac{x}{2} = x$$

This is called as the efficient cost function (Ulph 1996). The producer's strategic behavior is associated with the choice of ϕ. Any level other than the efficient level indicates a strategic behavior.

The production of good x generates pollution. Producers can reduce their emissions by choosing a level of costly abatement, a. The cost of abatement is equal to $\frac{a^2}{2}$. The reason producers need to spend some of their resources on costly abatement is that governments in each country set standards for the level of emissions generated. By choosing units suitably, the level of emissions that the producer generates will be equal to $e = x - a$ (it is being assumed that one unit of output generates one unit of emission). The welfare cost associated with the emissions is assumed to be $\frac{de^2}{2}$, where d is a positive constant. Governments determine the optimal emission levels, e and ε (emissions in foreign country), by maximizing their welfare functions associated with the production of this good, defined as total revenue minus total cost of production including cost of abatement minus total cost of pollution damage. We assume that the environmental damage is only local. Producers maximize their profit function, which is equal to total revenue minus total cost of production minus total cost of abatement, using the emission level determined by the government.

There are three stages of the game. In the first stage, governments choose the emission levels, e and ε, domestic and foreign respectively. Then the producers take the emission levels as given, and choose their R&D level, ϕ and ψ, domestic and foreign respectively. At the final stage the producers choose their output levels using the emission level set by the governments, and the R&D levels set in the second stage. The governments and producers can act strategically. In the present context, the governments are assumed to act strategically if they consider the impact of their emission levels on the output of the foreign producer. The producers act strategically by considering the side effects of their investment in R&D on the rival's output. When neither the government nor the producer is acting strategically, the government chooses the emission level ignoring the impact of its emission level on the foreign producer's output level, and the producer uses its efficient R&D level, $\phi = 2/x$ and $\psi = 2/y$, for domestic and foreign producers respectively.

Thus, four possible cases will be present pertaining to the behaviour of governments and producers in each country. These are (1) neither the government nor the producer is acting strategically, (2) only the government is acting strategically, (3) only the producer is acting strategically, and (4) both the government and the producer are acting strategically. These cases are valid for both countries. So, overall we can observe 16 different cases related to the behaviour of governments and producers in these two countries.

The profits of the domestic firm will be given by:

$$\pi_D = (A - x - y)x - x - 0.5(x - e)^2$$

In this expression production cost function is taken as the efficient cost function described above. If the firm chooses its R&D level strategically, the profit expression needs to be modified.

The welfare level of the domestic country is given by

$$W_D(e) = (A - x - y)x - x - 0.5(x - e)^2 - 0.5de^2$$

It is expressed as a function of the level of emissions, which is the government's choice variable. The welfare expression is equal to producer profits net of environmental disutility. The good is assumed to be sold in a third country market in order to abstract from consumer surplus calculations.

Similarly for the foreign country, the profit and welfare expressions will be given by:

$$\pi_F = (A - x - y)y - y - 0.5(y - \varepsilon)^2.$$

$$W_F(\varepsilon) = (A - x - y)y - y - 0.5(y - \varepsilon)^2 - 0.5d\varepsilon^2.$$

3. Welfare Analysis for Non-symmetric Cases

Governments may behave strategically or efficiently in determining their environmental policies; producers may behave strategically or optimally in choosing their R&D levels. Our main objective is to determine the Nash equilibrium of the strategic game between the countries. Identification of the Nash equilibrium requires the determination of the welfare levels (for the governments' behavior) and the profit levels (for the producers' behavior). Out of 16 possible set of behaviors, the four symmetric cases[1] have been studied by Ulph (1996). More explicitly, these cases are: 1) neither the governments nor the producers are acting strategically in both countries (NS—NS), 2) only governments are acting strategically(GS—GS), 3) only producers are acting strategically (PS—PS), and 4) both governments and producers are acting strategically (Both—Both). The remaining 12 hybrid cases will be examined in this paper. For each of the 16 cases, the welfare levels and producers' profits will be computed to determine the Nash equilibrium at the end. Welfare and profit expressions corresponding to the 4 symmetric cases will be computed by using the findings of Ulph (1996).

3.1 Non-strategic versus Only Foreign Government is Acting Strategically, (NS versus GS)

In this case the domestic government, domestic and foreign producers are acting non-strategically, and only the foreign government is acting strategically. Since the producers are not acting strategically in each country they choose their

[1]By symmetry we mean that governments and producers in both countries act in the same manner.

outputs by fixing the R&D parameter at the efficient level. Domestic firm's problem is as follows:

$$\max_{x}(A - x - y)x - x - 0.5(x - e)^2 \quad (4.1)$$

The producers take the emission level and the output of the rival as given. The first order condition for this maximization problem results in:

$$x = (A - 1 + e - y)/3 \quad (4.2)$$

Given the symmetry, foreign producer will have the following first order condition:

$$y = (A - 1 + \varepsilon - x)/3 \quad (4.3)$$

where ε is the emission level of the foreign country. Substituting (4.2) and (4.3) into each other we get:

$$x = \frac{(2A - 2 + 3e - \varepsilon)}{8} \quad (4.4)$$

$$y = \frac{(2A - 2 + 3\varepsilon - e)}{8} \quad (4.5)$$

In the first stage of the game, the domestic government chooses domestic emission level, e, by maximizing the welfare function (foreign emission level, ε, and the output level, y, are treated as given):

$$\max_{e} W(e) = (A - x - y)x - x - 0.5(x - e)^2 - 0.5de^2 \quad (4.6)$$

The first order condition is:

$$(A - 1 - y + e - 3x)\frac{\partial x}{\partial e} + x - e - de = 0$$

From equation (4.2), $(A - 1 - y + e - 3x) = 0$, so the first order condition becomes:

$$x - e = de \quad (4.7)$$

$$e = x/(1 + d) \quad (4.8)$$

Equation (4.7) represents a familiar outcome: the equality of marginal abatement cost (left side) and marginal damage cost (right side). Foreign government acts strategically by taking into consideration that both x and y depend on the level of emissions it chooses, ε. Foreign government maximizes

$$\max_{\varepsilon} W(\varepsilon) = (A - x - y)y - y - 0.5(y - \varepsilon)^2 - 0.5d\varepsilon^2 \quad (4.9)$$

The first order condition associated with the welfare maximization in (4.9) is equal to:

$$(A-1-y+\varepsilon-3x)\frac{\partial y}{\partial \varepsilon}-y\frac{\partial x}{\partial \varepsilon}+y-\varepsilon-d\varepsilon=0$$

$$\varepsilon(1+d)=y-y\frac{\partial x}{\partial \varepsilon} \tag{4.10}$$

From equation (4.4), we get $\frac{\partial x}{\partial \varepsilon}=-1/8$. Substituting this value into (4.10) gives us:

$$\varepsilon=\frac{9y}{8(1+d)} \tag{4.11}$$

We substitute the values for emission levels into equations (4.4) and (4.5) and get the following reaction functions:

$$x=\frac{16(A-1)(1+d)-9y}{64(1+d)-24}\quad,\quad y=\frac{16(A-1)(1+d)-8x}{64(1+d)-27}$$

Substituting one into the other gives us:

$$x=\frac{2(A-1)(1+d)\left[64(1+d)-36\right]}{(5+8d)\left[64\left(1+d\right)-27\right]-9} \tag{4.12}$$

$$y=\frac{2(A-1)(1+d)\left[64(1+d)-32\right]}{(5+8d)\left[64\left(1+d\right)-27\right]-9} \tag{4.13}$$

$$e=\frac{2\left(A-1\right)\left[64\left(1+d\right)-36\right]}{(5+8d)\left[64\left(1+d\right)-27\right]-9} \tag{4.14}$$

$$\varepsilon=\frac{9}{4}\frac{\left(A-1\right)\left[64\left(1+d\right)-32\right]}{(5+8d)\left[64\left(1+d\right)-27\right]-9} \tag{4.15}$$

Given these output and emission expressions, the corresponding welfare levels will be equal to:

$$W_D(e)=(A-x-y)x-x-0.5(x-e)^2-0.5de^2.$$

$$W_F(\varepsilon)=(A-x-y)y-y-0.5(y-\varepsilon)^2-0.5d\varepsilon^2.$$

where x, y, e and ε are given in equations (4.12)–(4.15).

3.2 Non-strategic versus Only Foreign Producer is Acting Strategically (NS versus PS)

Since the producer in the domestic country is not acting strategically, its profit maximization problem will result in the same reaction function in equation (4.2). Foreign producer is acting strategically, so it chooses the R&D level by maximizing its profits with respect to its R&D parameter, ψ in the second stage instead of choosing it through the minimization of total cost. Using a backwards solution approach, foreign producer determines its output level y as a function of the R&D level, ψ in the third stage:

$$\max_{y}(A - x - y)y - \psi\frac{y^2}{4} - 0.5(y - \varepsilon)^2 \tag{4.16}$$

The second term is the total cost function assuming that the level of R&D, ψ, has been determined in the second stage. The first order condition is:

$$y = \frac{2(A - x + \varepsilon)}{6 + \psi} \tag{4.17}$$

Solving the reaction functions in equations (4.2) and (4.17) together we get

$$x = \frac{(4 + \psi)A + 6e - 2\varepsilon - 6 - \psi}{3\psi + 16} \tag{4.18}$$

$$y = \frac{(12 + 2\psi)(2A + 1) - 12e + (20 + 3\psi)\varepsilon}{(3\psi + 16)(6 + \psi)} \tag{4.19}$$

At the second stage, foreign producer maximizes its profits with respect to the R&D level (ψ). This is where the strategic behavior of foreign producer becomes visible:

$$\max_{\psi}(A - x - y)y - \frac{\psi y^2}{4} - \frac{1}{\psi} - 0.5(y - \varepsilon)^2 \tag{4.20}$$

The first order condition corresponding to (4.20) is:

$$[A - x + \varepsilon + 0.5\,(6 + \psi)\,y]\,\frac{\partial y}{\partial \psi} - y\frac{\partial x}{\partial \psi} - \frac{y^2}{4} + \frac{1}{\psi^2} = 0$$

From equation (4.17), the first term in the above expression is equal to zero, and we get:

$$\frac{1}{\psi^2} = \frac{y^2}{4} + y\frac{\partial x}{\partial \psi} \tag{4.21}$$

From equation (4.18):

$$\frac{\partial x}{\partial \psi} = \frac{4A + 2 - 18e + 6\varepsilon}{(3\psi + 16)^2}$$

Since none of the governments acts strategically, as shown in the previous section, emission levels will be equal to $e = x/(1+d)$ and $\varepsilon = y/(1+d)$ for domestic and foreign countries, respectively. All combined we have 5 equations in 5 unknowns, namely x, y, e, ε and ψ. An analytic solution is not possible, and we solve this system by using numerical solution procedures. Then, welfare expressions can be obtained accordingly.

3.3 Non-strategic versus Both Foreign Government and Producer are Acting Strategically (NS versus Both)

The producer in the first country maximizes the profit function in (4.1); foreign producer behaves as in the previous section. The only difference with the previous case is the strategic behavior of the foreign government, which results in the emission level, $\varepsilon = \frac{y}{(1+d)}(1 - \frac{\partial x}{\partial \varepsilon})$ as in equation (4.10). It can be shown that $\frac{\partial x}{\partial \varepsilon} = \frac{-2}{(16+3\psi)}$ (from equation 4.18), and $\varepsilon = \frac{y(9+1.5\psi)}{(16+3\psi)}$. Then the welfare levels can be obtained by using numerical solution procedures.

3.4 Only Domestic Government and Foreign Producer are Acting Strategically (GS versus PS)

The reaction function for the domestic producer is same as the one in equation (4.2), and the reaction function for the foreign producer is given by equation (4.17). The strategic behavior of foreign producer requires that the R&D level satisfy the condition in equation (4.21). The domestic government behaves strategically, and thus, domestic emission level, e, will be determined by considering its impact on x, y, ϕ, and ψ. Maximization of domestic welfare under strategic behavior yields the following first order condition:

$$(A - 3x - y - 1 + e)\left(\frac{\partial x}{\partial e} + \frac{\partial x}{\partial \psi}\frac{\partial \psi}{\partial e}\right) - x\frac{\partial y}{\partial e} - x\frac{\partial y}{\partial \psi}\frac{\partial \psi}{\partial e} + x - e - de = 0 \quad (4.22)$$

From equation (4.2) the first term disappears, and (4.22) simplifies to:

$$e\,(1+d) = x\left(1 - \frac{\partial y}{\partial e} - \frac{\partial y}{\partial \psi}\frac{\partial \psi}{\partial e}\right) \quad (4.23)$$

We see that the deviation from the non-strategic rule is in two terms: $\frac{\partial y}{\partial e}$, the strategic incentive to influence the output of the foreign producer directly, and $\frac{\partial y}{\partial \psi}\frac{\partial \psi}{\partial e}$, the strategic incentive to influence the output of the foreign country indirectly, (through the foreign producer's investment in R&D).

Since foreign government does not act strategically, foreign emission level will be equal to $\varepsilon = y/(1+d)$. The R&D level for the foreign producer will be given by equation (4.21) above. It is not possible to find an analytic solution for this case; as in the previous case, we use numerical solution procedures to determine the corresponding welfare levels.

3.5 Domestic Government and Both Foreign Government and Producer are Acting Strategically (GS versus Both)

The domestic producer chooses its output level as in equation (4.2), and the foreign producer as in equation (4.17). The strategic behavior of foreign producer requires that the R&D level satisfy $\frac{1}{\psi^2} = \frac{y^2}{4} + y\frac{\partial x}{\partial \psi}$ (as in equation 4.21). Both governments behave strategically, and thus they consider the impact of emission levels on the output levels of the producers both directly and indirectly (through R&D parameter). Domestic emission levels, e, will be determined from equation (4.23). Foreign government sets its emission level by using $\varepsilon = \frac{y}{(1+d)}(1 - \frac{\partial x}{\partial \varepsilon})$. It can be shown that $\frac{\partial x}{\partial \varepsilon} = \frac{-2}{(16+3\psi)}$ (from equation 4.18), and so, $\varepsilon = \frac{y(18+3\psi)}{(16+3\psi)}$. We have 5 equations in 5 unknowns, namely x, y, e, ε and ψ. An analytic solution is not possible, and again, we solve this system by using numerical solution procedures. Then, welfare expressions can be obtained accordingly.

3.6 Domestic Producer, and Both Foreign Government and Producer are Acting Strategically (PS versus Both)

Producers in both countries will behave strategically; this will result in the following reaction functions:

$$x = \frac{2(A - y + e)}{6 + \phi} \tag{4.24}$$

$$y = \frac{2(A - x + \varepsilon)}{6 + \psi} \tag{4.25}$$

Solving these two reaction functions in (4.24) and (4.25) together, we get:

$$x = \lambda[(8 + 2\psi)A + 2(6 + \psi)e - 4\varepsilon] \tag{4.26}$$

$$y = \lambda[(8 + 2\phi)A + 2(6 + \phi)\varepsilon - 4e] \tag{4.27}$$

where $\lambda = 1/(32 + 6\phi + 6\psi + \phi\psi)$. λ is always positive since ϕ and ψ are always positive.

At the second stage producers maximize their profit functions with respect to their R&D levels (ϕ and ψ). This is where the strategic behavior of producers becomes visible. For the domestic firm, the R&D level is determined frrom:

$$\max_{\phi}(A - x - y)x - \frac{\phi x^2}{4} - \frac{1}{\phi} - 0.5(x - e)^2 \tag{4.28}$$

Equation (4.28) is slightly different from (4.1) in the inclusion of the R&D cost in the objective function; in the third stage of the game there is no need to include the R&D cost since R&D level will already be chosen.

The first order condition corresponding to (4.28) is:

$$[A - y + e + 0.5\,(6 + \phi)\,x]\,\frac{\partial x}{\partial \phi} - x\frac{\partial y}{\partial \phi} - \frac{x^2}{4} + \frac{1}{\phi^2} = 0$$

From equation (4.24), the first term in the above expression is equal to zero, and we get:

$$\frac{1}{\phi^2} = \frac{x^2}{4} + x\frac{\partial y}{\partial \phi} \tag{4.29}$$

From equation (4.27):

$$\frac{\partial y}{\partial \phi} = 2\lambda x$$

Putting this in equation (4.29), we get

$$\frac{1}{\phi^2} = \frac{x^2}{4}\,(1 + 8\lambda) \tag{4.30}$$

Similarly for foreign producer:

$$\frac{1}{\psi^2} = \frac{y^2}{4}\,(1 + 8\lambda) \tag{4.31}$$

Domestic government behaves non-strategically, and will set the emission level efficiently at $e = \frac{x}{1+d}$.

The foreign government behaves strategically and maximizes:

$$(A - x - y)y - \frac{\psi y^2}{4} - \frac{1}{\psi} - 0.5(y - \varepsilon)^2 - 0.5d\varepsilon^2 \tag{4.32}$$

The first order condition associated with (4.32) will be:

$$\begin{aligned}&[(A - x + \varepsilon) - 0.5y(6 + \psi)]\left(\frac{\partial y}{\partial \varepsilon} + \frac{\partial y}{\partial \psi}\frac{\partial \psi}{\partial \varepsilon} + \frac{\partial y}{\partial \phi}\frac{\partial \phi}{\partial \varepsilon}\right)\\ &+\left(-y\frac{\partial x}{\partial \psi} - \frac{y^2}{4} + \frac{1}{\psi^2}\right)\frac{\partial \psi}{\partial \varepsilon} + y\left[1 - \frac{\partial x}{\partial \varepsilon} - \frac{\partial x}{\partial \phi}\frac{\partial \phi}{\partial \varepsilon}\right] - \varepsilon - d\varepsilon = 0 \end{aligned}\tag{4.33}$$

Using equations (4.25) and (4.31), the above first order condition simplifies to:

$$\varepsilon(1 + d) = y\left[1 - \frac{\partial x}{\partial \varepsilon} - \frac{\partial x}{\partial \phi}\frac{\partial \phi}{\partial \varepsilon}\right] \tag{4.34}$$

From equations (4.26) and (4.27), we get:

$$\frac{\partial x}{\partial \varepsilon} = -4\lambda, \quad \frac{\partial x}{\partial \phi} = -\lambda x(6 + \psi), \quad \frac{\partial \phi}{\partial \varepsilon} = \frac{4\lambda\phi}{x}$$

Using these, we rewrite (4.34) as:

$$\varepsilon = \frac{y\left[1 + 4\lambda + 4\lambda^2(6 + \psi)\phi\right]}{1 + d} \tag{4.35}$$

Substituting equations (4.30) and (4.35) into (4.26) and (4.27), we get:

$$x = \frac{f(1 - g') - uf'}{(1 - g)(1 - g') - uv} \tag{4.36}$$

$$y = \frac{f'(1 - g) - vf}{(1 - g)(1 - g') - uv} \tag{4.37}$$

where

$$f = (8 + 2\psi)A\lambda, \quad g = \frac{2\lambda(6 + \psi)}{1 + d}, \quad u = \frac{4K\lambda}{1 + d},$$

$$f' = (8 + 2\phi)\, A\lambda, \quad g' = \frac{2\lambda\,(6+\phi)\,K}{1+d}, \quad v = \frac{4\lambda}{1+d}$$

$$K = 1 + 4\lambda + 4\lambda^2\,(6+\psi)\,\phi.$$

It is not possible to solve for the output, emission and R&D levels analytically, and we obtain these variables and corresponding welfare levels by using numerical solution procedures.

Given the identical country assumption, the welfare levels corresponding to the next six cases where the behaviors of foreign and domestic countries are interchanged can easily be obtained by substituting domestic variables for foreign variables and vice versa. Welfare levels corresponding to the four symmetric cases analyzed in Ulph (1996) will be taken from Ulph directly. Now, we start to analyze the Nash equilibrium.

4. Nash Equilibrium

In the previous section we have identified the welfare levels of both countries for all possible set of behaviors. Now, by using them we will determine the Nash equilibrium of the game played by the governments and producers. Firstly, governments choose their environmental policy, then producers choose their R&D and output levels. Given that producers are rivals in a third country market, environmental policy and R&D levels may not be set at their efficient levels to generate competitive advantage. Thus, an interesting issue is whether producers and governments will act strategically or efficiently.

Welfare levels computed in the previous section can not be compared analytically; thus we are forced to assume numerical values for the basic parameters, A and d. As in Ulph (1996), we let $A = 10$ and $d = 1$. We have also considered other possible values for A and d, and the results were qualitatively similar. The pay-off table constructed from the welfare expressions obtained in the previous section for the 16 possible cases is given in Table 4.1.

Producers' actions will be based on their profit levels, and Table 4.2 presents the profit levels for the domestic and foreign producers for each of the 16 cases. We should note that in both tables, there are only two actions for the players: strategic or non-strategic behavior; however, as there are two different player types, the returns to action of one type (for example, government) depends on what other type (producer) is doing. Therefore, for the sake of notational simplicity, we have placed four "actions" for each player. We will now determine the Nash equilibrium of the simultaneous and sequential move versions of the

Table 4.1. Welfare Levels for Domestic and Foreign Countries.

	NS	GS only	PS only	Both
NS	8,505	8,532	8,149	7,968
	8,505	8,593	8,297	8,337
GS only	8,593	8,252	8,177	8,017
	8,532	8,252	8,103	8,143
PS only	8,297	8,103	8,423	8.014
	8,149	8,177	8,423	8.231
Both	8,337	8,143	8.231	8.070
	7,968	8,017	8.014	8.070

In each cell, top value belongs to domestic country, and the bottom value belongs to foreign country. Strategies for domestic (foreign) government are given in the rows (columns).

game represented by the pay-off tables in tables 4.1 and 4.2.

Table 4.2. Profits of Domestic and Foreign Producers

	NS	GS only	PS only	Both
NS	9,315	9,311	8,964	8,765
	9,315	9,611	9,153	9,452
GS only	9,611	9,330	9,262	9,082
	9,311	9,330	8,944	9,231
PS only	9,153	8,944	9,248	8.839
	8,964	9,262	9,248	9.324
Both	9,452	9,231	9.324	9.136
	8,765	9,082	8.839	9.136

In each cell, top value belongs to domestic producer, and the bottom value belongs to foreign producer. Strategies for domestic (foreign) producer are given in the rows (columns).

4.1 Simultaneous Move Game

In the simultaneous move game, first governments then producers will move simultaneously. Note that the pay-off tables are different for governments (Table 4.1) and producers (Table 4.2). If a Nash equilibrium exists, then the corresponding set of actions for producers and governments must match; this means that the Nash equilibrium corresponding to the pay-off tables above must indicate the same actions. Thus, we will determine the Nash outcome for both pay-off tables separately, and check if there is a common set actions.

From Table 4.1, there is multiple equilibrium for governments' game: NS—GS, GS—NS, PS—PS, and Both—Both (first action belongs to domestic government and the second one to foreign government). From Table 4.2, we see that producers' game has also multiple equilibrium: GS—GS, and Both—Both (first action belongs to domestic producer and the second one to foreign producer). Note that GS means that producer is not acting strategically, and Both refers to the case that both producer and government are acting strategically as described in the previous section. Both-Both is the only common action pair, so it is the unique Nash equilibrium for the simultaneous move game. Leaving the discussion of this result to the conclusion, we now move to the sequential move game.

4.2 Sequential Move Game

There may be alternative definitions of sequential move game in our context (as there are two different types of agents). Nevertheless, we will only consider the one in which one of the governments moves first, and then the other government acts after observing this action. Producers are assumed to move simultaneously. Sequential behavior in the setting of environmental policy is not a strong assumption as it has been observed in reality much often especially in the developed and developing country context.

We will assume that domestic government moves first. From table 4.1, it is clearly seen that the highest return that the domestic government can secure comes from GS-only action, i.e. domestic (leader) country will prefer to act strategically. The foreign government, the follower, then will act efficiently (non-strategically—NS action). Under these actions by governments, the pay-off table for producers (Table 4.2) reduces to the following subgame pay-off table.

Table 4.3. Sequential-move Game. Producers' Profits

	NS		PS only	
GS only	9,611	9,311	9,262	8,944
Both	9,452	8,765	9.324	8.839

In each cell, left value belongs to domestic producer, and the right value belongs to foreign producer. Strategies for domestic (foreign) producer are given in the rows (columns).

As seen clearly, there are two equilibria: GS—NS and Both—PS. These indicate that either both producers behave strategically or none of them behaves strategically. Producers get higher profits if they do not act strategically, and choose their R&D levels efficiently. Then, we can argue that Both—PS equilibrium is inferior to GS—NS equilibrium; thus in the sequential move game only the leader government will behave strategically, and all other players will behave non-strategically (efficiently).

4.3 Emissions

Now, in this partial equilibrium-imperfect-competition framework, we can present the impact of trade on the environment. This will be done by computing the emission values for each of the 16 cases. The resulting emissions are given in Table 4.4 below.

Table 4.4. Emission Levels for Domestic and Foreign Countries

	NS	GS only	PS only	Both
NS	1,2857	1,2699	1,2766	1,2624
	1,2857	1,4442	1,3083	1,4931
GS only	1,4442	1, 4727	1,4735	1,4594
	1,2699	1,4727	1,2969	1,4755
PS only	1,3083	1,2969	1,2979	1,2841
	1,2766	1,4735	1,2979	1,4782
Both	1,4931	1,4755	1,4782	1,4599
	1,2624	1,4594	1,2841	1,4599

In each cell, top value belongs to domestic country, and the bottom value belongs to foreign country.
Strategies for domestic (foreign) country are given in the rows (columns).

In the simultaneous move game, the equilibrium emission levels (corresponding to Both—Both) will be much higher than the efficient levels (corresponding to NS—NS) due to strategic behavior of governments and producers. Note that, producers' strategic behavior partially offsets the increase in emission levels associated with strategic behavior of governments (see the emissions corresponding to GS—GS). As producers invest more into the R&D when they act strategically, higher R&D level reduces emission levels. In the sequential move game, the emission levels will be higher in the domestic country (corresponding to the equilibrium behavior GS—NS). GS-NS equilibrium is also superior to Both—PS equilibrium with respect to resulting emissions. It is clearly observed that strategic behavior of governments increases the level of emissions, which is the cost of providing competitive advantage to their producers. Finally, emission levels are lowest when government and producer in one country behave efficiently (i.e. non-strategically) and the government and producer in the other country behave strategically (see the cells corresponding to NS versus Both or

Both versus NS in table 4.4).

5. Summary

Will international trade lead to relaxation of environmental policies? Will the incentives of governments for acting strategically in setting their environmental policies be higher if producers competing in international markets behave strategically? What will be the Nash equilibrium of the strategic game between governments and producers in the presence of international trade and environmental regulations? How does the equilibrium change when we move from a simultaneous move game to a sequential move game?

In this chapter, we have analyzed the impact of trade liberalization in an imperfectly competitive market. In a standard model used in earlier studies in this literature, we analyze the behaviors of governments and producers related to environmental policy making and R&D investment, respectively. Governments may act strategically by recognizing that the output of the foreign producer depends on the emission level it sets, and thus, it may distort its environmental policy in order to generate competitive advantage to its producer. Producers may act strategically by adjusting the level of investment in R&D by considering the associated impact on the output of the rival producer. Our main objective is to determine the Nash equilibrium of this game including governments and producers in the presence of environmental regulations, international trade and innovation. Surprisingly, this has not been reported before.

We show that there is a unique Nash equilibrium in the simultaneous move game, and both producers and governments behave strategically. Both countries end up with a lower welfare and higher emission level than the case in which none behaves strategically. This is like a prisoner's-dilemma-type outcome. In the sequential move game, two equilibria exist, but one dominates the other in terms of both the welfare and emission levels. In the equilibrium, first-moving (leader) government acts strategically, the follower government will not act strategically, and none of the producers will behave strategically. The first-moving country gets higher welfare than the follower country. Similarly, producer in the first-moving country gets higher profits than the producer of the follower country. Furthermore, as expected, emission level is higher in the first-moving country because of the strategic behavior.

Our explicit analysis of the welfare levels and profits not only reveals the equilibrium behaviors but also puts doubt on some of the earlier findings in this literature. For example, Ulph (1996) states that allowing governments to act strategically increases the incentive for producers' to act strategically. As seen in the pay-off tables we have presented, producers will prefer to behave non-strategically (i.e. efficiently) when governments act strategically. Producers will prefer to act strategically only if the other producer is acting strategically. Finally, the highest welfare levels correspond to the cases in which only one government acts strategically and the other one efficiently, but this set of behaviors can not be sustained in the simultaneous-move game.

In this chapter, we have studied only the symmetric case, i.e., both countries were identical. In future work, it will be interesting to release this assumption, and extend our analysis to non-symmetric cases. A first step towards such an extension is shown in Alpay and Karaman (2002).

Chapter 5

GENERAL EQUILIBRIUM MODELS OF TRADE AND ENVIRONMENT

General equilibrium models of trade and environment have been employed with different focus in the literature. As opposed to partial equilibrium models, general equilibrium approach makes it possible to analyze the direct and indirect effects of any changes in the economy, therefore it is more useful for policy development. Nevertheless, the complexity introduced by the incorporation of multiple markets sometimes make the results less intuitive and interesting. In most cases, special functional forms need to be assumed on both the demand and supply side of the economy to obtain solutions at the desired degree of usefullness.

Usually two-country, two-good, two-input Heckscher-Ohlin type trade models have been employed with the incorporation of environment in different ways. In general, modellings have been done at an aggregated level (for example, usually a representative consumer assumption is employed, thus the potential impact of any policy, here environmental, changes on different, say, income groups are assumed away.) Different types of modelling assumptions mentioned in the first part of partial equilibrium chapters are also relevant here. Environment is either modelled as an input into the production process or as a by-product of production; environmental degradation is either consumption or production-based; governments apply environmental policies (optimally or strategically) or environmental resources are non-regulated, etc.

In a general equilibrium modelling the usual impact of environmental policies on the overall welfare level, environmental degradation can be identified together with their impact on output and input price levels, supply and demand of all goods. Hence, comparison of different environmental policies in terms of their efficiency and distributive effects can be done more meaningfully. In

S. Alpay (ed.), Trade and The Environment, 51–66.

this chapter, we will briefly overview the literature, and summarize the modelling assumptions and findings of some of the earlier papers, and thus give the reader the opportunity to assess the set-up, operation and solution of the general equilibrium approaches to the analysis of environmental consequences of trade liberalization.

1. Literature Overview

The general equilibrium (GE) modelling approaches to trade and environment have been employed in a number of papers. One can mention Pethig (1976) and Asako (1979) as the pioneers of models analyzing trade and environment in a GE model. These papers consider two-country, two-good models with pollution being treated as an input (in Pethig), and as a joint product (in Asoko). Environmental policy is carried out with usual environmental tools (either emissions standards or pollution taxes) in Pethig versus product tax-subsidy policies in Asoko. Their common result is that the country exporting the pollution intensive good may loose from trade when the environmental cost of higher pollution arising from increased production exceeds the trade gains. Secondly, as environmental regulations become more restrictive in one country, the relative factor endowments will be affected and the country with the relatively laxer regulations will have comparative advantage in pollution-intensive good. Pethig also shows that environmental policy changes may have impact on the location of production facilities.

Mcguire (1982) studies differences between coordinated and uncoordinated (unilateral) pollution regulation in a slightly modified Heckscher-Ohlin model. As usual, the model contains two countries, and two commodities. Changes from the standard model includes (i) pollution as input in addition to the usual two inputs, capital and labor, in the production of one good, and (ii) the introduction of environmental regulation. Firstly, the impact of regulation in autarky is analyzed with respect to returns to factors of production, and it is shown that the factor which is intensively used in the non-regulated industry gains relatively more. Then, the analysis is carried into a trading equilibrium.

In the abscence of environmental regulations, factor prices will be equalized through trade. Two types of regulations are distinguished: coordinated versus uncoordinated. In a coordinated regulation, the price of pollution input remains the same in both countries. In an uncoordinated or unilateral regulation, the equality of factor prices is lost. The overall impact depends on whether the world prices are affected by the introduction of environmental regulations. If the country is a small country then the changes in factor returns will mimic the ones in autarky, and thus, returns to the factor used intensively in the regulated sector will decrease. If the country is a large one, then the world prices will be affected by the environmental regulations. McGuire shows that factor returns will remain the same in the regulating country; however, in the non-regulating country, the factor used intensively in the regulated sector will gain relative to the other factor. Finally, if factor mobility is introduced into the model, then complete specialization in production will be seen in at least one country if there is unilateral regulation (due to factor price equalization). Thus, if the pollution is of a transnational type with complete symmetry between countries, the unilateral regulation will have no impact on the level of pollution; the only impact will be that total pollution now will be produced in the non-regulating country. Nevertheless, if the pollution intensity differs between countries, and for example, if the non-regulating country requires higher pollution input per unit of output, then pollution level will increase in both countries paradoxically following the unilateral environmental regulation.

Merrifield (1988) models transnational pollution explicitly in a general equilibrium trade framework with international mobility of capital. An interesting side of this paper is the use of abatement equipment standards as the regulatory instrument in place of usual types such as pollution taxes, emission standards, or tradeable permits. Otherwise, the model is similar to the earlier two-country, two-good Heckscher-Ohlin models. Merrifield analyzes the model under complete specialization assumption. Production of both goods generates pollution. The productivity of the capital input is assumed to depend on the level of transnational pollution negatively. Consumers' welfare expressions only include consumption of both goods, and do not incorporate a pollution disutility.

Similar to McGuire (1982), the paradoxical result that pollution increases with the introduction of a new tax on the output of the polluting industries has been shown in more detail by using the example of US-Canada acid deposition problem. The critical factor here is the differences in the endowments of factors of production across countries. If the US imposes a production tax on its output, then the relative prices will be altered, and this will affect the returns to factors. Given that capital input is internationally mobile, some capital will move out of US into Canada. Because the size of the US capital stock is much higher, the relative sizes of capital changes will be very different between US and Canada. Therefore, the reduction in the output of US will be relatively smaller than the increase in the Canadian output. It is quite possible that the post-regulation decline in emissions in US could be overwhelmed by the increase in the Canadian production, and the regulation may result in higher emissions. A similar policy in Canada will more likely achieve the objective of reducing the overall emissions due to differences in the sizes of the US and Canadian economy.

Merrifield considers abatement equipment standard as the main regulatory instrument in his model. If instead of production tax, US government carries out the environmental policy by raising abatement equipment standards, market responses will be different. There will be direct and indirect effects. Increases in the abatement equipment standards directly reduce US emissions; moreover, movement of capital and labor into the abatement sector from the production of the polluting good generates further reductions. As the US capital market becomes tight initially and capital prices increase, some capital moves into the US from Canada given the international mobility of capital. This reduces Canadian emissions. The indirect effect arises from the increased productivity of capital resulting from the reduced emissions. With higher productivity of capital, the production of goods increase both in US and Canada, which will generate new emissions. The overall impact depends on the differences between US and Canada in (i) the elasticity of substitution of capital for labor in the production of final goods, (ii) the capital intensiveness of abatement equipment production, and (iii) elasticity of capital productivity with respect to the pollution flow. Estimated values of these parameters from real data for US and Canada fall into the ranges for which an overall decline in emissions will be warranted after increases in abatement equipment standard.

We now move to a more detailed analysis of some recent papers incorporating trade and environment in a general equilibrium modelling. Our objective is to introduce different general equilibrium models to the interested readers in more detail, giving them the chance to see alternative set-up possibilities.

2. Differences in Property Rights

Chichilnisky (1994b) identifies the impact of freer trade on the use of natural resources when some countries lack clearly defined property rights. She considers a general equilibrium model with two goods, two inputs and two countries, similar to a standard Heckscher-Ohlin one except that the supply of the inputs are price-dependent. One of the inputs is chosen as environmental resource (E), which is extracted from a resource pool by either capital or subsistence labor. The novelty of the Chichilnisky model is that one country (South) has ill-defined property rights regime as compared to the other (North). This key feature is introduced into the model through the supply functions of the environmental resource; at any given price the supply of E in the South with ill-defined common property rights is higher than that of North which has well-defined private property rights:

$$E = \alpha \frac{P_E}{q} + E^0 \qquad (5.1)$$

where P_E is the price of the environmental resource, and q is the opportunity cost of the input used in extracting E. Here, the coefficient α represents the property rights regime for the pool from which E is extracted, and differs between North and South (α for South is larger than that of the North). North and South differ only with respect to this aspect. Similarly the supply of the other input of the model, capital, is also price-dependent:

$$K = \beta r + \bar{K} \qquad (5.2)$$

where r is the rental rate on capital, and β and $\bar{K}$ are constants. On the production side, fixed-proportions technology is assumed in the production of goods A, and B with

$$A = E^A/a_2 = K^A/c_2 \qquad (5.3)$$
$$B = E^B/a_1 = K^B/c_1 \qquad (5.4)$$

On the demand side, Chichilnisky uses a specific utility function:

$$\begin{aligned} U(A,B) &= B + k \qquad && if \;\; A \geq A^{D*} \\ U(A,B) &= B + \gamma A && otherwise \end{aligned} \qquad (5.5)$$

where $\gamma = k/A^{D*}$, and k is a positive number. If the relative price of good B is more than $1/\gamma$, then the demand for good A will be satiated at A^{D*}, which will simplify the analysis of the model. Chichilnisky mentions that this very specific utility function can be extended to more general functional forms at the cost of more complex computations. Then, the model is solved by using equilibrium conditions: (i) supply-demand equilibrium and (ii) trade equilibrium. Factor prices, allocation of factors across sectors, goods prices and production of goods are then determined by solving the system of equations in a general equilibrium framework. Firstly, the paper shows the existence of an equilibrium price ratio for final goods, p. Then the factor prices are determined from

$$P_A = a_1 P_E + c_1 r \qquad (5.6)$$
$$P_B = a_2 P_E + c_2 r \qquad (5.7)$$

Given P_E and r, one can solve for E and K from equations (5.1 and 5.2) above respectively. Then the rest of the model can easily be solved. She shows that two countries with identical technologies, endowments, and preferences will trade with one another if one has ill-defined property rights on her environmental resources. The country with ill-defined property rights already overuses the environment as a factor of production; trade with a country with well-defined property rights will further increase the overuse of resources, and will make the present misallocation worse. Although trade equalizes the prices of traded goods and factors of production, this does not improve resource allocation. In the resulting world economy, resources are underpriced and there is overproduction in the South, and overconsumption in the North. The implication of this result is very vital as many environmental resources are unregulated common property in developing countries. Rain forests, which are used for timber or destroyed to give way to the production and export of cash crops like coffee, sugar, and palm oil, constitute an important example.

Another interesting result of the paper is that traditional policies designed to overcome this common resource problem may not be adequate, and in fact, may make the situation worse. To show this, Chichilnisky (1994b) assumes the existence of subsistence labor, which is only employed in the extraction of environmental resource in the South. Introduction of a unit tax for the use of environmental resource may not deter overextraction. Taxes can force lower income harvesters to work harder and extract more resources to keep up with their pre-tax consumption behavior. As a result, taxes can lead to more rather than less extraction of the resource (after-tax price of environmental resource is lower, and this leads to a drop in the opportunity cost of the labor of the subsistence workers. As the resource prices go down, subsistence workers put more effort and hence more resources are extracted, rendering the environmental policy ineffective.). Thus, policies aimed at correcting the property rights problem should be preferred instead of resource tax in the presence of subsistence workers. If capital is used in the extraction of resources instead of subsistence labor, then taxation will result in lower resource use; however, in most of the developing countries resource extraction is done by subsistence labor, and hence the results of the Chichilnisky paper have important practical implications.

Chichilnisky also mentions the implications of her findings regarding the higly debated statements of Lawrence Summers first originated in a World Bank memorandum in 1991, and then cited in an editorial in the Economist in 1992. Summers argues that 'the World Bank should be encouraging more migration of the dirty industries to the less developed countries'. In interpreting Summers' argument, Khan (1996) states that 'the desirability of free trade leads inevitably and inexorably to the desirability of free trade in pollution'. Chichilnisky suggests that the apparent comparative advantages may not be actual comparative advantages. As in her paper, the driving factor of trade between countries may have nothing to do with relative abundance of resources, differences in productivity or preferences and/or lower factor prices, but instead, it may be associated with differences in institutional factors such as the property rights regime. In case of poorly defined property rights regimes, which is the case in most of the developing countries, her model shows that these countries will be exporting environmentally intensive goods at levels more than efficient, and at prices less than the social cost. Therefore, the efficiency argument behind Summers' statement may not be warranted.

3. Differences in Human Capital

Copeland and Taylor (1994) investigate the impact of differences in human capital on the environmental consequences of trade liberalization. Main features of the paper can be grouped under three headings. Firstly, the model is of a North-South type to capture the impact of income differences. Income disparity will generate different incentives for enacting environmental policies: with higher income, demand for environmental quality will rise, and that may trigger introduction of higher environmental regulations[1]. If the only difference between countries is the level of environmental policies[2] (thus, the sole determinant of trade pattern is the differences in the environmental policies), then it is interesting to determine whether such trade is inherently pollution-generating or not.

Secondly, the model establishes the link between income levels and environmental policy by assuming that benevolent planning agents in each country set the environmental policy by applying the optimal rule, that is to say, by equating the pollution taxes equal to the marginal damage generated by emissions. Moreover, the policy is endogenously determined, and thus the comparison between autarky and trade is done in the presence of optimal environmental policies in each case.

Thirdly, as opposed to most of the papers in these literature using a two-good model, they use a many-good general equilibrium modelling a la Dornsbusch, Fischer and Samuelson (1977). The advantages of this set-up is many-fold: the full impact of environmental policies on factor markets, incomes and trade flows can be shown; additionally, the composition-of-output impact of free trade on both the national and world pollution levels will be identified under the assumption of different pollution intensities across industries. Now, we introduce the

[1]This is known as technique effect in this literature. It will be explained in more detail together with composition and scale effect below.
[2]In their case the level of pollution taxes.

model used in Copeland and Taylor (1994) in more technical terms.

As in Dornsbusch, Fischer and Samuelson (1977), the model includes a continuum of consumption goods, indexed by $z \in [0,1]$. Pollution is assumed to be local[3], and is modelled as an input to the production process together with labor. The production function is of Cobb-Douglas type:

$$y(d,l;z) = \begin{cases} l^{1-\alpha(z)} d^{\alpha(z)} & \text{if} \quad d \leq \lambda l \\ 0 & \text{if} \quad d > \lambda l \end{cases}$$

where l is the labor input and d is pollution discharge; λ is a positive number, and it indicates limits to substitution possibilities between labor and pollution; $\alpha(z)$ represents the share of pollution charges in the overall cost of producing good z and is between 0 and 1. Countries (North and South) are distinguished from each other only with respect to differences in the human capital. Northern human capital, h is assumed to be higher than the Southern one, h^*. The effective labor supply in the North, represented by $A(h)L$ is higher than that of South, $A(h^*)L$.

On the demand side, the utility function incorporates both the consumption of commodities, and damage from pollution:

$$U = \int_0^1 b(z) ln[x(z)] dz - \frac{\beta(L,\rho) D^\gamma}{\gamma} \tag{5.8}$$

where $x(z)$ is the consumption, and $b(z)$ is the budget share of good z; D is the total amount of pollution generated, and $\beta(L,\rho)$ shows the utility cost factor of pollution with respect to population, and population density. In most of the analysis, β is treated as constant due to identical-country assumption except for the level of human capital. Under the assumption that all consumers own one unit of labor, and pollution tax revenues are distributed back to the consumers equally, Copeland and Taylor obtain the indirect utility function corresponding to equation (5.8) as:

$$V = \int_0^1 b(z) ln[b(z)] dz - \int_0^1 b(z) ln[p(z)] dz + ln(\frac{I}{L}) - \frac{\beta D^\gamma}{\gamma} \tag{5.9}$$

where I and p refer to income and prices respectively. Note that income is given by $I = wL + \tau D$ where τ is the pollution tax rate, and w is the wage rate.

[3]Extension to transnational pollution is shown below.

Governments set their environmental policies optimally by maximizing the indirect utility function of the representative consumer in (5.9) with respect to pollution tax, τ, which results in:

$$\tau = \beta D^{\gamma-1} I \tag{5.10}$$

Pattern of trade depends on the relative magnitudes of unit cost functions of a given commodity across countries. If the variables corresponding to South are indicated by an asterisk, commodity z will produced in the North if

$$c(w, \tau; h, z) \leq c(w^*, \tau^*; h^*, z) \tag{5.11}$$

Using the production function, and applying cost minimization to get the open form of cost functions, equation (5.11) can be expressed in a compact form:

$$\frac{w}{w^*} \leq \frac{A(h)}{A(h^*)} \left(\frac{\tau^*}{\tau}\right)^{\frac{\alpha(z)}{1-\alpha(z)}} \tag{5.12}$$

Using the last equation together with trade balance condition, Copeland and Taylor obtain their first result regarding the existence of trade equilibrium between South and North. Under the assumption that North has sufficiently higher human capital as compared to the South, it will have relatively high income, and thus, Northern pollution tax will be higher than Southern one $\tau > \tau^*$. This will force pollution-intensive industries to move to South, and relatively cleaner industries will locate in the North. In the trading equilibrium, North will produce all goods with index $z < \bar{z}$, and South produces all goods with index $z > \bar{z}$. Note that goods are indexed initially so that lower indexes indicate low pollution-intensive industries, and vice versa.

The main objective of the paper is to analyze the impact of trade on the pollution level in the North, South and world. This has been done by comparing the autarky pollution levels with pollution levels after trade. Firstly, the channels through which trade affects the pollution levels are introduced. The total amount of pollution D in the North after trade can be expressed as:

$$D = \frac{I\theta(\bar{z})}{\tau\phi(\bar{z})} \tag{5.13}$$

where $\theta(\bar{z}) = \int_0^{\bar{z}} \alpha(z)b(z)dz$ is the share of northern pollution charges in the world income; $\phi(\bar{z}) = \int_0^{\bar{z}} b(z)dz$ is the share of world spending on the goods produced in the North (see Copeland and Taylor 1994, page 765). Taking the total differentiation of (5.13), one gets

$$dD = \frac{\theta(\bar{z})}{\tau\phi(\bar{z})}dI + (-\frac{I\theta(\bar{z})}{\tau^2\phi(\bar{z})})d\tau + \left(\frac{Ib(\bar{z})}{\tau\phi(\bar{z})^2}\int_0^{\bar{z}}[\alpha(z) - \alpha(\bar{z})]b(z)dz\right)d\bar{z} \tag{5.14}$$

The first term in the right-hand side of equation (5.14) is known as the *scale effect*. The scale effect shows the changes in the level of pollution generated by a change in the level of economic activity when the technique of production and the composition of final output are kept constant. It is positive for both North and the South[4].

The second term in (5.14) is known as the *technique effect*. The technique effect shows the changes in the pollution level due to a switch to less pollution-intensive methods of production when income (or scale of the economy) and the composition of final output are kept constant. It is negative for both North and the South.

Finally, the last term in (5.14) is known as the *composition effect*. The composition effect reflects changes in the level of pollution arising from changes in the composition of final output when the technique of production and the income level (scale of the economy) are kept constant. It is positive for both North and the South. The composition of final output depends on the trade pattern, which is identified by the position of index $\bar{z}$. If $\bar{z}$ increases, the bundle of goods produced in the North (South) increases (decreases). As the new goods added to the Northern production are relatively more pollution-intensive[5], pollution rises in the North. Similarly, the industries leaving the South are those which are relatively less pollution-intensive than the ones that remain in the South, therefore, average pollution intensity increases in the South as well.

In the set-up developed by Copeland and Taylor (1994), the composition effect always dominates the scale and the technique effect (i.e. the difference between the scale and the technique effects as they are in opposite signs). The main result of the paper then can be summarized as follows: If a trading equilibrium exists, trade will always lower the pollution level in the North, and increase the pollution level in the South; the decline in the northern pollution can not overcome the increase in the southern pollution, and thus trade increases world pollution under the assumption that factor prices are not equalized. If factor prices are made equal through trade, then world pollution level will remain

[4]Southern counterpart of equation (5.14) can be derived easily.

[5]Note that larger values of commodity index z indicates higher pollution intensity.

unchanged. We can briefly mention some key steps of the proof of this main result. Copeland and Taylor first show that the pollution demand and supply functions depend on both the relative input price ratio[6], $\frac{\tau}{w_e}$, and the effective human capital. As the human capital levels are different between North and the South, it is shown that in autarky, pollution is relatively scarce input in the North, and effective labor is relatively scarce input in the South. These differences form the basis of the trade, and determine the trade pattern: higher pollution-intensive industries stay in the South, and relatively clean industries move to the North. As stated above, the sign of the composition effect determines the overall impact. The paper then identifies the impact of moving resources from a relatively clean industry (say z') to a dirty one (say z'') in the South, and from a relatively dirty industry (z'') to a relatively clean one (z') in the North. The impact of such a reallocation on the world pollution is shown to be equal to

$$\Delta D_W = \Delta D_N + \Delta D_S = \left\{ \frac{\alpha(z'')}{1-\alpha(z'')} - \frac{\alpha(z')}{1-\alpha(z')} \right\} \left\{ \frac{w_e^*}{\tau^*} - \frac{w_e}{\tau} \right\} \quad (5.15)$$

where ΔD_W, ΔD_N, and ΔD_S show the changes in the level of pollution in the world, North and South, respectively. Since the model assumes that $\alpha'(z) > 0$ and since $\frac{w_e}{\tau} < \frac{w_e^*}{\tau^*}$ due to differences in the human capital, $\Delta D_W > 0$. If factor price equalization holds, it is clear from equation (5.15) that the world pollution level will remain the same after trade (the last term will be equal to zero).

Finally some comparative statics have been performed with respect to changes in human capital levels and income transfer. If the gap between Northern and Southern human capitals increases (decreases) then the increase in pollution after trade will be larger (smaller). Therefore, the environmental impact of trade liberalization can be improved with increases in the economic development of the developing countries. As Copeland and Taylor state "... as rich countries get richer, world pollution increases, but as poor countries get richer, world pollution falls". The gap between human capitals can be reduced by different policies, one of which is the income transfer from developed countries to the developing countries. income transfer from North to South reduces world pollution as well as Southern pollution, but increases Northern pollution. This is due to the increase in pollution taxes in the South (with most pollution-intensive industries) and a decline in the Northern pollution taxes as a result of transfer.

[6]The effective wage rate, w_e, is equal to regular wage rate multiplied by effective human capital, $A(h)$.

3.1 Transnational Pollution

Copeland and Taylor (1995) extend the previous model in several respects. Firstly, they allow for transnational pollution; secondly they consider an arbitrary number of countries, and look at the cases with a large number of countries (no terms-of-trade effect) and with a small number of countries (terms-of-trade effect). Thirdly, the regulatory instrument is changed to tradeable emissions permits from pollution taxes. Finally, equilibria in which factor-price-equalization (FPE) holds, and specialized equilibria are analyzed explicitly. We will not go into the details of the set-up, which is essentially derived from the Copeland and Taylor (1994) model described above, but instead summarize the main findings of the paper.

Copeland and Taylor (1995) show that if human capital levels differ substantially across countries, then a movement from autarky to free trade raises world pollution; if they are similar, world pollution does not rise with free trade. The driving force behind these two different results is whether factor prices, including emission permit prices, are equalized or not through trade; when countries are similar, factor prices will be made equal after trade, and the decline in the pollution level of human-capital rich country will be exactly off-set by the increase in the human-capital scarce country. These results seem to be similar to the ones in Copeland and Taylor (1994), but not exactly. As the present model assumes transnational pollution, increases in the Southern pollution directly affect Northern countries as well. In fact, after trade, each Northern country becomes worse-off as compared to autarky whereas all Southern countries gain from trade. Hence, Northern countries will be willing to link free trade agreements to environmental agreements to prevent the increase in the transnational pollution generated in the South. Such incentives are not present for Southern countries.

In a specialized equilibrium, free trade in goods raises world pollution. In such an equilibrium, allowing for international trade in pollution permits can counteract the rise in global pollution. When permit prices are not equal, dirtier industries will shift to those countries with lower permit prices (pollution-haven effect), and this movement rises global pollution. If pollution permit prices get equalized, then the pollution-haven effect will be eliminated, and global pollution will fall relative to the pure goods trade level.

Thirdly untied international income transfers, in an FPE equilibrium, will lower the recipient's pollution but raise the donor's pollution, and thus may have no effect on global pollution as well as on prices, and surprisingly on welfare levels of either country. On the other hand, income transfers tied directly to pollution reduction can be welfare enhancing. This last result underlines the potential importance of income effects both in analyzing global pollution reform, and in determining how international trade affects the global environment.

Finally, Copeland and Taylor (1995) consider the case where terms-of-trade effects can not be ignored. If aggregate demand and supply for pollution do not depend on world distribution of income and if factor price equalization holds, then terms-of-trade effect has no impact on the global pollution level; that is to say, autarky and free-trade pollution levels will be equal, and only the composition of pollution will change across countries (more Southern pollution and less Northern pollution). This is quite surprising as it implies that world pollution level remains unchanged even if countries distort their environmental policies to gain terms-of-trade advantage. This result is very much dependent upon the assumption that global pollution is pure public bad, and affects all countries equally.

4. Environmental Protection as a Composite Commodity

General equilibrium models that we have seen above modelled pollution either as a by-product of production or an input into the production process. Khan (1996) takes a different route, and considers environmental protection as a public good produced by the government, and paid for by the sectors according to their pollution intensities. In his model, environmental cleaning-up is provided by the goverment 'in recognition of the fact that responsibility for environmental degradation can not be assigned to one industry and the production of a single commodity'. The cost of production of the environmental good then must be shared by the polluters according to their pollution intensities.

Khan's model contains a rural sector, X_r, a disaggregated urban sector producing n different commodities, X_{iu}, and a non-traded intermediate composite commodity, X_g devoted to environmental protection. Production in the rural sector does not generate environmental degradation, and environmental degradation is due to urban production with different pollution intensities across sectors. The economy has two factors of production: capital (K) and labor (L); they are intersectorally mobile, but internationally immobile. Production functions are formally given by

$$X_{iu} = F_{iu}(L_{iu}, K_{iu}, X_g) \qquad (i = 1, 2, 3, ..., n) \qquad (5.16)$$

$$X_r = F_r(L_r, K_r) \qquad (5.17)$$

$$X_g = F_g(L_g, K_g) \qquad (5.18)$$

X_g is produced by the government and it essentially represents the amount of resources reserved for environmental protection. It is pure public good, and as seen in equation (5.16) all urban sectors use identical amount of X_g. Government finances the production of environmental good a la Lindahl taxation, and charges q_{iu} for the ith urban sector for its pollution. The wage rate is w and rental rate of capital is R. Khan (1996) describes the equilibrium of his model as follows:

"An equilibrium is a strictly positive tuple constituted by $\left((X_{iu}^*, L_{iu}^*, K_{iu}^*, q_{iu}^*)_{i=1}^{n}, (X_r^*, L_r^*, K_r), (X_g^*, L_g^*, K_g^*)\right)$ and by the pair (w^*, R^*) such that for all i:

1 $(L_{iu}^*, K_{iu}^*, X_g^*)$ maximizes $p_{iu}F_{iu}(L_{iu}, K_{iu}, X_g) - w^*L_{iu} - R^*K_{iu} - q_{iu}X_g$ and $X_{iu}^* = F_{iu}(L_{iu}^*, K_{iu}^*, X_g^*)$;

2 (L_r^*, K_r^*) maximizes $p_rF_r(L_r, K_r) - w^*L_r - R^*K_r$ and $X_{iu}^* = F_r(L_r^*, K_r^*)$;

3 (L_g^*, K_g^*) maximizes $(\sum_{i=1}^{n} q_{iu})F_g(L_g, K_g) - w^*L_g - R^*K_g$ and $X_g^* = F_g(L_g^*, K_g^*)$;

4 $\sum_{i=1}^{n} L_{iu}^* + L_r^* + L_g^* = L$;

5 $\sum_{i=1}^{n} K_{iu}^* + K_r^* + K_g^* = K$"

where L and K represent total labor and capital supply, respectively. All production functions are assumed to be continuously differentiable and constant-returns-to-scale. The paper then shows analogues of the standard trade theorems such as Stolper-Samuelson, and Rybczynski theorems under this set-up. We will only summarize those results that have relevance to the environmental impact of trade liberalization. The country is assumed to be a small country (a price-taker), and terms-of-trade considerations disappear. In this set-up, we will be interested in the determination of changes in the production of environmental commodity, X_g, as a result of trade.

Under the small country assumption, the country will have no impact on the world prices, and thus it will export those goods for which the world price is higher than the domestic price, and it will import those goods for which the world price is lower than the domestic price. Thus as a result of trade, domestic prices will be made equal to world prices (in case of exports (imports) there will be an increase (decrease) in the domestic prices). In order to obtain the impact of openness to international markets on the production of environmental commodity, Khan uses the framework developed by Jones (1965) incorporating input-output coefficients as a function of factor prices. Nevertheless, given the multisectoral nature of his model, the resulting expressions do not present intuitively clear conclusions. As an alternative approach, he mentions the use of gross national product (GNP) function in order to demonstrate the well-behavedness of the price-output responses, then he suggests the use of unit cost functions to identify the changes in the production of environmental composite good as a result of trade. The paper does not explicitly show the derivations. We will do that here briefly. Given the constant returns to scale assumption and by using the Shephard's Lemma, the input demand function for X_g will be given by

$$X_g = X_{iu} \frac{\partial C_{iu}(w, R, q_{iu})}{\partial q_{iu}} \qquad \text{for all } i = 1, 2, ...n \qquad (5.19)$$

where $C_{iu}(w, R, q_{iu})$ is the unit cost function. Trade will have impact on the prices of traded commodities; thus by using (5.19), we can identify the impact of urban commodity price changes on the production of environmental good. By taking the derivative of both sides with respect to the price of the ith urban commodity, one gets

$$\frac{\partial X_g}{\partial p_{iu}} = \frac{\partial X_{iu}}{\partial p_{iu}} \frac{\partial C_{iu}(w, R, q_{iu})}{\partial q_{iu}} + X_{iu} \left\{ \frac{\partial^2 C_{iu}}{\partial w \partial q_{iu}} \frac{\partial w}{\partial p_{iu}} + \frac{\partial^2 C_{iu}}{\partial R \partial q_{iu}} \frac{\partial R}{\partial p_{iu}} + \frac{\partial^2 C_{iu}}{\partial q_{iu}^2} \frac{\partial q_{iu}}{\partial p_{iu}} \right\} \qquad (5.20)$$

Khan's analysis shows that if the urban sector is capital intensive relative to the rural sector, then an increase in the price of jth urban output leads to an increase in the unit charge imposed on jth urban industry if and only if the collection of urban sectors with the exclusion of jth one is also capital intensive relative to the rural sector. Hence, if we assume (i) capital intensive urban sectors (in all different definitions of capital intensity) and (ii) suitable substitutability relationship among capital, labor and environmental good, then all terms in (5.20) can be shown to be positive. Thus, a positive relationship between the supply of environmental commodity and price of the ith urban output can be obtained. In such a case, as a result of trade liberalization, prices will increase in the export sectors, and thus more environmental good, X_g, will be produced; in the import sectors, domestic prices will go down and the supply of X_g will get reduced. The overall impact depends on the relative magnitudes of these opposing changes. In the extreme cases, if only exports (imports) are present, then the supply of environmental protection will in(de)crease.

Finally we would like to close this section with a classic application of Rybczynski theorem. If the urban sector is capital intensive relative to the rural sector, an increase in the endowment of capital input generates an identical increase in all urban commodities as well as in environmental good (capital intensity in this set-up is defined as follows: Urban sector is capital intensive relative to the rural sector if $\frac{K_g + \sum_{i=1}^{n} K_{iu}}{L_g + \sum_{i=1}^{n} L_{iu}} > \frac{K_r}{L_r}$).

5. Summary

In this chapter, we have presented a survey of the general equilibrium modelling approaches to the theoretical analysis of the environmental consequences of trade liberalization. We do not see a strong consensus on neither the modelling structure nor the outcome of the analysis. Some studies use a two-good two-country traditional Heckscher-Ohlin type models, some use multi-sector and some others use a continuum of goods models. In some papers, the focus is on the status of property-rights (ill-defined or well-defined), and in some others the attention is shifted to human capital differences to capture the impact of income on environmental policy making. Regarding the impact of trade on environment, the implications of our partial equilibrium analysis in the previous chapters are reinforced: the final impact depends on country-specific factors. In addition to factors indicated in the partial equilibrium chapter, it is shown that factors such as property-rights and human capital also have to be taken into consideration. Several new dimensions regarding the impact of trade on the environment have been introduced in this chapter as well. These include the incorporation of transnational pollution, and income transfers (tied and untied to environmental regulations). Finally, we mention two recent studies which present extensive analysis of interactions between trade and the environment. Copeland and Taylor (2001) set out a general equilibrium pollution and trade framework which contains a canonical pollution haven model as well as the standard Heckscher-Ohlin-Samuelson factor endowments model as special cases. Kraus (2000) approaches the interactions between trade and the environment from the opposite angle, and studies the question whether tariffs may serve as environmental policy instruments. Although these eco-tariffs are used to restrict trade as opposed to trade liberalization, the modelling approaches in Kraus are valuable extensions of standard Heckscher-Ohlin-Samuelson model, and can be used in the context of trade liberalization as well by suitable transformation.

Chapter 6

EMPIRICAL EVIDENCES ON THE ENVIRONMENTAL IMPACT OF TRADE LIBERALIZATION

1. Introduction

In the previous chapters, we have seen a spectrum of theoretical approaches on the environmental impact of free trade. Many different modelling frameworks from partial equilibrium ones to many-good, many-country general equilibrium models have been utilized. Different combinations of environmental policy making have been studied. Overall, results vary among many different papers; some studies indicate that trade liberalization worsens environmental conditions, whereas some others provide evidences for improved environmental conditions as a result of freer trade.

The impact of trade liberalization on the environment is dependent on a number of country-specific factors: the origin of environmental degradation (consumption versus production), the comparative advantage of the country (whether it is importer or exporter of the pollution-intensive goods), the size of the country, small or large (whether the country can affect world price level, so whether terms-of-trade changes are relavant or not), etc. Therefore, the same liberalization policy may result in opposite effects in countries which differ with respect to the mentioned characteristics.

As a general guideline, the impact of trade liberalization on the environment can be decomposed into three parts as introduced by Grossman and Krueger (1991): the scale effect, which represents the changes in the size of the economic activities; the composition effect, showing the changes in the bundle of goods being produced; and the technique effect, representing the changes in the production technology, mainly adoption of cleaner technology. In general, scale effect tends to increase the amount of environmental degradation,

S. Alpay (ed.), Trade and The Environment, 67–84.

technique effect lowers it; the sign of composition effect depends on the comparative advantages of the countries, and thus, may be positive in some cases and negative in others. Thus, trade liberalization will generate positive environmental consequences if technique effect outweighs the scale and composition effects (in case of a country with comparative advantage in dirty industries) or if technique effect together with composition effect outweighs the scale effect (in case of a country with comparative advantage in clean industries).

Given the ambiguity in the thereotical approaches, empirical studies become much more valuable. We can talk about three general approaches in the empirical literature: (i) Past data analysis; (ii) Policy simulations in Computable General Equilibrium (CGE) models, and (iii) Case studies from individual countries. Most studies used CGE modelling as it permits the analysis of many potential policy options from limited liberalization to full–scale openness to international markets. In the following sections we will summarize some of the important empirical papers in each group. We will not be able to mention all studies in the empirical literature in order to expose some leading studies in more detail.

2. Past Data Analysis

Lopez (1997) presents both theoretical and empirical analysis regarding the impact of trade liberalization on the environment. His principal purpose is to estimate the value of environmental resources as factors of agricultural production, and to measure the potential effects of various economy-wide policies (mainly trade liberalization policies) on agricultural income by explicitly accounting for environmental effects. The analysis has been carried out in a two-sector general equilibrium framework. Along with the conventional factors of production, like labor and capital in the urban sector, agricultural sector also uses an environmental input, called biomass in the paper. The empirical analysis is done on Ghana, where the government has been taking important steps towards liberalizing the economy since 1983. Lopez (1997) uses a unique data set for an area in Western Ghana. These data include household survey information from 12 villages for the periods 1988-1989 with remote sensing data on the area under closed forest, area under natural bush (fallow area), and density of biomass in the fallow area. The detailed survey includes data on production, land use, employment, use of conventional factors of production, and demographic characteristics.

First, a general theoretical approach is presented, and it is shown that if the environmental externality (as reflected in land cultivation decisions) is fully internalized, trade liberalization will unambiguously increase national income;

however, if land cultivation decisions are socially inefficient, the net effect of trade liberalization is, in general, ambiguous. His estimations show that the impact of deepening of trade liberalization on biomass depletion (a form of environmental deterioration) is quite significant: further losses of biomass on the order of 2.5 to 4 percent follow the trade liberalization. The net income elasticity associated with trade liberalization fluctuates between -0.002 and -0.0078 for various simulations executed for a wide range of non-farm labor demand elasticity. Thus, further trade liberalization generates a fall, albeit small, in the national income due to its detrimental effects on the environment.

Dean (1998) develops a structural simultaneous-equation model which incorporates the static and dynamic interactions between trade and the environment. The basics of her model are: National income, Y, is modelled as a function of trade restrictions, conventional factors of production (capital and labor), and the level of emissions, E, (which proxies the use of environment); equilibrium level of emissions is endogenously determined; the derived demand for E is a function of emissions charge, income and the share of pollution intensive goods in real output. She considers a small economy which produces two types of goods, dirty and clean. The utility function of the representative consumer is a function of consumption of these two goods and the level of emissions. Solving her model, Dean (1998) obtains the following simultaneous equation model in the growth of income and the emissions:

$$\hat{Y} = \alpha_E \hat{E} + \alpha_F \hat{F} + \hat{A}$$

$$\hat{E} = \left(\frac{\epsilon_\tau}{|\theta|}\right)(\hat{p}_D - \hat{p}_C) - \epsilon_Y \hat{Y}$$

where α_E and α_F represent the share of inputs in total output, F represents the aggregator stock of the conventional factors of production; A is a measure of increased openness; ϵ_τ is supply elasticity of emissions with respect emission charges; ϵ_y is income elasticity of demand for clean environment; $|\theta|$ represents the difference between the share of emissions in the unit cost of the produced goods; and finally $(\hat{p}_D - \hat{p}_C)$ shows the terms of trade.

In this system of equations trade liberalization affects the growth of emissions in two different ways. A reduction in trade restrictions will raise the relative price of dirty goods (second equation), leading to increased specialization in dirty goods, and so a rise in the growth of emissions. On the other hand, lower levels of trade protection will increase factor productivity, and thereby income (from first equation). This increase in income will reduce the growth of emissions as the willingness of individuals to supply the environment as a factor of production decreases. In brief, Dean's model reveals that trade liberalization directly aggravates environmental damage via its influence on the terms of trade, but indirectly mitigates it via its effect on income growth.

Dean estimated the model described above by using a two-stage least squares method on pooled Chinese provincial data on industrial water pollution emissions over the period 1987–1995 (obtained from the World Bank). The choice of water pollution emission data is partly due to data availability, but also due to the fact that Chinese firms do pay levies on water pollution. The critical part in the empirical analysis is the estimation of the trade restrictiveness at the aggregate level, and she used the black market premium as a proxy for this. The estimated results appear to validate her hypothesis. An increase of 1% in the relative price of net exports lead to an increase in the growth of emissions by 0.08% (this is the direct impact of trade liberalization on pollution). At the same time, a 1% reduction in trade restrictiveness produces an increase in income growth, which in turn causes a decline in the growth of emissions by 0.0088%. She names this indirect impact of trade liberalization as technique effect, which is due to the positive association between income growth and demand for clean environment.

One of the most comprehensive studies in this literature is by Antweiler et al. (2001). This paper starts with setting out a theory of how openness to international markets affects pollution levels to assess the environmental consequences of international trade. In their model, Antweiler et al. assume a small open economy that has two primary factors (labor and capital) and produces, in competitive markets, two final goods, one of which is capital-intensive and generates pollution as a by-product. There are N consumers with preferences over consumption goods and clean environment. It is assumed that government sets a pollution tax as an increasing function of optimal tax (which is determined by the maximization of the sum of utilities of consumers). This kind of tax policy not only allows for the possibility that government behavior varies across countries (due to political systems, for instance), but also allows pollution policy to respond endogenously to changing economic conditions.

As the empirical model is based on the outcomes of the theoretical analysis, a brief introduction on the theoretical findings will be useful. If we compare countries with similar incomes and scale, we expect to find that openness will be associated with higher pollution in countries with a comparative advantage in the polluting good, and with lower pollution in countries with a comparative advantage in the clean good. This result tells us that looking for a simple correlation between openness and environmental quality is unlikely to be fruitful, and instead, we have to focus on the link between openness, comparative advantage, and pollution. In the model of Antweiler et al. (2001), comparative advantage is determined by the interplay of relative factor endowments and differences in pollution policy (which is mainly determined by per capita income). With this in the background, they show that if a country is sufficiently rich then the pollution haven motive for trade will eventually outweigh factor endowment considerations, and this country will export the clean good in trade (thus, openness will be associated with lower pollution). On the other hand, if a country is sufficiently capital abundant then the factor endowment basis for trade will eventually outweigh any pollution haven motive for trade and this country will export the dirty good (thus, openness will be associated with higher pollution).

In the second part of the paper, empirical investigations have been carried out. They use the Global Environment Monitoring System data on sulfur dioxide emissions. Their data set consists of 2621 observations from 293 observation sites located in 109 cities representing 44 countries spanning the years 1971-1996. The majority of the data on economic variables (GDP per capita, population, capital stock per worker, trade intensity) come from the Penn World Tables. The estimated model is:

$$\begin{aligned} z_{ijkt} &= \beta_0 + \beta_1 GDP_{jkt} + \beta_2 KL_{kt} + \beta_3 KL_{kt}^2 + \beta_4 I_{kt} + \beta_5 I_{kt}^2 + \beta_6 R_{ijk} \\ &\quad + \beta_7 B_{ijk} + \beta_8 M_{jkt}^T + \beta_9 M_{jkt}^P + \beta_{10} O_{kt} + \beta_{11} O_{kt} RKL_{kt} \\ &\quad + \beta_{12} O_{kt} RKL_{kt}^2 + \beta_{13} O_{kt} RI_{kt} + \beta_{14} O_{kt} RI_{kt}^2 + \epsilon_{ijkt} \end{aligned}$$

where i is index for site, j for city, k for country, and t for time; z represents pollution concentrations; GDP is measured by real GDP per unit area[1] (obtained by multiplying city population density with country GDP per person); KL is capital to labor ratio; I is one period lagged three year moving average of GDP per capita; R is a dummy indicating site ijk is in rural location; B is a dummy for suburban location; M^T is average temperature; M^P is the variation in precipitation; O is measured by the ratio of exports and imports to GDP (an indicator of openness); $ORKL$ terms are interactions of openness with capital labor ratio; and ORI terms represent the interactions of openness with income. For the indicator of openness, one of the key variables of the model, other measures such as black market premium, average tariffs, average quota and Sachs and Warner dummy are also considered in the estimation stage.

The initial estimation results of Antweiler et al. do not provide evidence for a significant relationship between pollution and different indicators of openness. This is not surprising as the impact of openness on pollution concentrations in

[1]Antweiler et al (2001) explain the need for such a measure as follows: "To explain *concentrations of pollution* we need a measure of scale reflecting the *concentration of economic activity* within the same geographical area."

their set-up should only reflect the induced composition effect, which varies with country characteristics. For some countries it will be positive, and for some others it will be negative, and thus on average, no significant effect may be observed. When they condition the degree of openness on country characteristics (such as capital abundance and income), the estimated impact of openness on pollution concentrations turns out to be negative and highly significant. This indicates that higher openness results in lower sulfur dioxide emissions. The calculated trade intensity elasticity, which measures the predicted change in pollution concentrations for a 1% change in the ratio of exports plus imports to GDP, is -0.869. They also compute the associated scale, technique, and composition effects. The estimated values for the scale, technique and composition elasticities calculated at the sample means are 0.193, (-1.611), and 1.135 respectively (all are significant). These values conform to the implications of the theory. Taken together, the full impact of further openness to international trade, through scale, techniques and composition effects, will be a reduction in sulfur dioxide concentrations for an average country. They conclude that these estimations yield a somewhat "surprising" conclusion: freer trade appears to be good for the environment.

An interesting part of the empirical analysis pertains to the environmental Kuznets curve (EKC) hypothesis[2], which will be introduced in the next chapter. They try to distinguish between the environmental consequences of income changes resulting from openness versus capital accumulation or technological progress. Assuming a constant capital share in GDP of $1/3$, they show that growth induced by capital-accumulation only, will increase the level of pollution even after considering the impact of income on the environmental policy. This result together with the previous one (the negative impact of openness on pollution levels) indicates that "trade liberalization plus capital accumulation is far less environmentally friendly than trade liberalization alone" (see Copeland and Taylor 2001). Moreover, empirical tests of EKC hypothesis need to pay more attention to the source of the economic growth. The inverted-U type relationship may be indicating different sources of income growth at different levels of economic development.

Antweiler et al. (2001) seminal study presents a detailed exposition of both the theoretical and empirical assessment of the impact of freer trade on the environment. It has made important contributions to the understanding of this subject. Their framework should be utilized in deepening our understanding on this matter. For instance, sulfur dioxide is only one of the many different

[2]Briefly, EKC hypothesis suggests an inverted-U type relationship between pollution and income.

forms of environmental pollution. Other forms of environmental degradation should be incorporated into their empirical analysis. Although it will be a very demanding task, a further work which can incorporate an index of environmental quality, composing of many different environmental indicators as in Ekins (1997), will be a worthwhile attempt.

3. Computable General Equilibrium Models

Many studies have attempted to uncover the impact of trade on the environment through computable general equilibrium models. Empirical studies using CGE models differed from each other in several respects. Most notably, we can mention differences regarding the coverage of the countries and the type of the simulation policies considered. Some concentrated on a single country (see, for example, Dessus and Bussolo 1998, and Strutt and Anderson 1999), some on worldwide (Ferrantino and Linkins 1999, and Cole et al. 1998). Some studied the impact of liberalization policies implied by Uruguay Round (Ferrantino and Linkins 1999, and Cole et al. 1998), some considered integration to regional trade blocks, such as NAFTA, (Beghin et al. 1999), and some others examined complete liberalization (zero tariffs) scenarios (Lee and Roland-Holst 1997). We now present the details regarding some of these papers.

3.1 Single-Country CGE Models

Several very recent papers analyze the issue empirically in a Computable General Equilibrium (CGE) model. Lee and Roland-Holst (1997) assessed the linkage between trade and the environment in a two-country CGE model (for Indonesia and Japan) by removing Indonesia's nominal tariffs on all imports. This unilateral trade liberalization by Indonesia would increase the ratio of emission levels to real output for almost all major pollution categories (total percentage changes for the 10 different pollutants included in the study range from 0.51 to 3.73 for biological oxygen demand-BOD and lead, respectively.). If tariff removal policy is combined with a uniform effluent tax policy (as opposed to an export tax or sector specific taxes), the twin objectives of welfare enhancement and environmental quality improvement appear to be feasible from the simulation studies they have carried out. This result is supportive of the main conclusion of this survey: trade liberalization accompanied by environmental considerations may very well be welfare enhancing.

Beghin et al. (1997) analyze the linkages between growth, trade and the environment in Mexican agriculture with an empirical economy-wide model through the investigation of trade liberalization, environmental policy reform, and their coordination. Outward orientation induces pollution growth in some agricultural sectors. Their main result is that more liberal trade combined with targeted effluent taxes can achieve significant environmental mitigation and efficiency gains, but with the implication of a contraction of most agricultural sectors.

Beghin et al. (1999) investigate the linkages between trade integration, environmental degradation and public health for Chile. In a 72-sector CGE model incorporating monitoring functions for 13 effluent categories and a variety of mortality and morbidity indicators, they show that unilateral opening to world markets induces substantial worsening of pollution emissions and expansion of resource-based sectors. NAFTA integration is shown to be environmentally benign in terms of pollution emissions, and in fact, it reduces emissions in several pollutants. This decline in effluents is achieved through strong composition effects in production, outweighing the scale expansion induced by NAFTA; however, these decreases in emissions do not translate into major gains for urban health. Simulations for integration via MERCOSUR[3] case result in adverse effects on the environment and on urban morbidity and mortality. Finally, taxing air pollutants and simultaneously reducing trade distortions (by maintaining revenue neutrality) induces net welfare gains from both reduced health damages and increased efficiency, a double-dividend-type result.

In a recursive dynamic CGE model for Costa Rica, Dessus and Bussolo (1998) study a quantitative assessment of the interdependencies of trade liberalization and environmental policies. Similar to previous CGE papers, they demonstrate that trade liberalization induces a risk of specialization in dirty activities; however, free trade combined with appropriate effluent taxes reduces emissions significantly. Abler et al. (1999) also examine the environmental impacts of trade liberalization in Costa Rica in a CGE model. This model includes eight environmental indicators covering deforestation, pesticides, overfishing, hazardous wastes, inorganic wastes, organic wastes, greenhouse gases, and air pollution. As opposed to most other studies, Abler et al. (1999) incorporate possible technology changes after trade liberalization into their analysis. They find out that the impact of trade liberalization on the environmental indicators are generally negative, but small or moderate in magnitude.

In another study on Indonesia, Strutt and Anderson (1999) present simulation results on the environmental impact of two liberalization policies: the Uruguay

[3]MERCOSUR is a regional treaty for trade liberalization among Argentina, Brazil, Paraguay and Uruguay.

Round and most-favored-nation liberalization by Asian-Pacific Economic Cooperation (APEC). The paper is also valuable in that it incorporates the likely impact of new technology adoption on emissions. By performing a simulation in which Indonesia's income growth is assumed to increase by an extra 0.5 percent per year (a conservative estimate) as a result of trade liberalization, they show that by 2020 air pollution would be 12 to 15 percent greater and water pollution 6 to 12 percent greater following the liberalization. They suggest that the policy-makers have to make sure that the environmental reforms (especially in pollution-intensive or resource-extracting sectors) are put in place before severe environmental degradation occurs in order to make trade reform welfare improving.

3.2 Multi-Country CGE Models

Ferrantino and Linkins (1999) estimate the potential changes in toxic industrial emissions arising from two trade liberalization experiments conducted using computable general equilibrium (CGE) modelling. The study is very comprehensive in terms of its coverage of countries and sectors. The aggregated model consists of 10 regions representing the developed world (separated into four as North America(NAM), Japan(JPN), Australia–New Zealand–Norway–Iceland–Switzerland(ANE) and European Union (EU)), and developing world (grouped under six different regions namely Latin America (LAM), Africa–Middle East–South Asia (ROW), China-Hong Kong (CHK), South Korea–Singapore–Taiwan (SEA), and transition economies including former Soviet Union, Eastern Europe, North Korea (EIT)). The sectors considered in the study are leather, pulp and paper, chemicals, iron–steel, and non-ferrous metals, which generate very high toxic emissions, and textiles, and apparel which generate relatively moderate emissions.

Two trade liberalization scenarios are considered: (i) trade liberalization arising from Uruguay Round, and (ii) a "zero–for–zero" agreement that will eliminate all tariffs in manufactures. As the employed model is not a dynamic one, the first scenario represents what the effects of liberalization would have been if the full implementation had occurred in 1992.

Production, trade and protection data come from the Global Trade Analysis Project (GTAP) database. Liberalization experiments were analyzed by conducting simulations using the GTAP CGE model. On the production side, this model assumes perfect competition, constant returns to scale technology with internationally immobile factors of production (land, labor and capital); at the industry level, constant elasticity of substitution functions are used for the value added and composite nests. On the demand side, representative consumer modelling is used with the utilization of a constant difference of elasticities (CDE) functional form for private household demands.

Ferrantino and Linkins first estimate the changes in the output levels in each sector after the liberalization policies have been carried out, and then calculate the associated changes in the toxic emissions. The simulation results show that local percentage changes in toxic emissions will vary from -3.43 (in China and Hong Kong) to 1.84 (in Southeast Asia) for the first liberalization policy. The world emission for this case will remain almost unchanged. In the zero–for–zero liberalization case, regional toxic emissions vary between -7.74 (in China and Hong Kong) and 1.40 percent (in Korea, Singapore, and Taiwan); global toxic emissions will decline this time, although by a very small margin (0.18 %). Detailed results are shown in table 6.1. For the Uruguay Round experiment, the main impact of trade liberalization is a compositional change in the emission levels across regions including developing countries with no significant changes in developed ones. In China-Hong Kong emissions will decline by 3.4 percent; emissions in South-East Asia and South Korea-Taiwan-Singapore increase by 1.8 percent and 1 percent respectively. In the zero-for-zero experiment, the changes are more pronounced. This time China-Hong Kong experience an emission reduction of 7.7 percent; emissions in South-East Asia and Africa–Middle East–South Asia decline by 2.5 percent and 2.89 percent, respectively. Changes in emission levels in the developed countries are also more significant in this scenario; the largest impact is seen in Australia–New Zealand–Norway–Iceland–Switzerland with a fall of 2.3 percent.

Changes in emission levels are calculated under the assumption that U.S. emissions coefficients apply to all regions given the lack of data for the emission coefficients across regions. This assumption is highly questionable. Especially, as there is substantial evidence for higher toxic intensity of industries in developing countries, estimated changes in the global emissions are not very reliable for either of the two trade liberalization scenarios. Nevertheless, Ferrantino and Linkins indicate that regional changes are still valuable as they rely only on the assumption that differences in toxic intensities *across industries* are negligible *across regions*.

In the second part of their analysis, they try to address this problem by introducing regional technology differences. Two different estimates of emission intensities have been used: (i) emission intensities in developing regions are three times larger than in developed regions; (ii) emission intensities are computed as a function of per capita income by using environmental Kuznets curve (EKC) hypothesis implied in Grossman and Krueger (1995). EKC hypothesis

Table 6.1. Local Percentage Changes in Toxic Emissions (without technology effects)

Region	Uruguay Round	Manufacturing zero-for-zero
ANE	-0.19	-2.31
JPN	-0.01	0.54
KST	1.00	1.40
SEA	1.84	-2.50
CHK	-3.43	-7.74
ROW	-0.44	-2.89
NAM	-0.03	-0.06
LAM	-0.04	0.12
EU	0.07	0.18
EIT	0.57	0.69
World (assuming US technology)	-0.003	-0.18

Source: Ferrantino and Linkins (1999).

indicates an inverted U-type relationship between per capita income and emissions[4], and the regional emission coefficients are computed in accordance with the findings in Grossman and Krueger. Ferrantino and Linkins show that in both cases where the toxic intensity differs between developed and developing countries, trade liberalization generates higher environmental benefits as compared to the case under US emission coefficient assumption. This positive impact becomes more pronounced in the full liberalization policy as compared to Uruguay Round policy.

In the last part, Ferrrantino and Linkins incorporate the income induced technological changes. They state that "Trade liberalization in general induces improvements in living standards. If higher living standards are associated with reductions in the emissions coefficient, then trade liberalization may improve environmental quality through a second channel: liberalization raises incomes, and higher incomes lead to improved technology and reduced emissions".

They assume that all technological improvements are realized in developing regions. Firstly, the authors estimate the impact of the two experiments on different income measures, and identify which developing regions will experience significant income effects after trade liberalization. They find that in case of Uruguay Round liberalization, China-Hong Kong, South-East Asia and South Korea-Taiwan-Singapore will have significant increases in income,

[4]It will be explored in more detail in the next chapter.

and in case of complete-removal of manufacturing tariffs South-East Asia and South Korea-Taiwan-Singapore will experience positive income effects. As expected, incorporation of income-induced technological changes makes positive environmental benefits associated with trade liberalization stronger. In fact, in the case of South Korea-Taiwan-Singapore, the adverse environmental impact of both liberalization policies turns into a positive impact. Global emissions get reduced in more significant terms in both the Uruguay Round experiment and full manufacturing liberalization experiment.

In conclusion, Ferrantino and Linkins (1999) find that trade liberalization and environmental protection are complementary on the global scale. They explain this by the historical practice of granting high protection to high emission industries in developing countries, and the shift of dirty production from developing to developed countries following the liberalization. Additionally, improvements in income due to trade liberalization may lead to greater political demand for environmental regulation and/or greater investments in clean technologies. Ferrantino and Linkins also mention limitations of their study. Firstly, these results are only applicable to toxic emissions from manufacturing sector and can not be extended to other environmental problems such as deforestation, biodiversity etc. Additionally, the usual criticisms against CGE models should be remembered. Finally, not a limitation but a suggestion, this study, being one of the most comprehensive ones due to its coverage of sectors and countries, can be improved by the incorporation of a dynamic framework.

Another study regarding the environmental impact of trade liberalization a la Uruguay Round is done by Cole et al. (1998). This study considers five different pollutants, namely nitrogen oxides, sulphur dioxide, carbon monoxide, suspended particulate matter and carbon dioxide, as indicators of environmental quality. The empirical model includes nine countries/regions: EU, USA, Japan, China, East Asia, South Asia, Africa, South America and Eastern Europe. Cole et al. (1998) estimate the environmental impact of Uruguay Round liberalization through the explicit calculations of the composition effect, the scale effect and the technique effect mentioned above.

Two different scenarios have been considered: (i) the environmental impact in 1990, if the full effects of the Uruguay Round (UR) were in place, and (ii) the environmental impact in 2000, if the additional increase in income growth due to UR is assumed to be 0.5 percent, and stays same for 10 years upto 2000. In each scenario, the authors estimate the level of emissions for each pollutant with and without the Uruguay Round and record the difference (emissions under UR less of emissions without UR). Both the aggregate changes and the decomposition of this aggregate into composition and combined scale-technique effects are calculated for each scenario.

Cole et al. utilise the computable general equilibrium estimates of the output and income effects of the Uruguay Round from Francois et al. (1995); these are then converted into emission levels by multiplying with the sectoral pollution intensities from Hettige et al. (1995). Hettige et al. (1995) estimate pollution intensities for US industries; Cole et al. generates pollution intensities for other regions/countries by using the estimated environmental Kuznets curves for these regions/countries.

With respect to composition of output, Francois et al. (1995) estimate that most heavy industry sectors will expand in developed regions/countries; therefore, the composition effect in all of developed economies for all pollutants (except carbon dioxide) is positive; on the other hand, in developing regions composition effect is mostly negative with some exceptions (namely, South America and Eastern Europe).

The combined scale and technique effect is always negative (reduction in emissions) for all developed regions in case of pollutants sulphur dioxide, carbon monoxide and suspended particulate matter. In case of nitrogen oxides, and carbon dioxide, developed regions experience emission increases (the only exception is a decline in nitrogen oxides for USA in the 2000 scenario). For all pollutants, the combined scale and technique effect is almost always positive (higher emissions) in case of developing regions/countries (with the exception of Eastern Europe in some cases).

The aggregate impact has been calculated for the four pollutants (excluding carbon dioxide). In case of sulphur dioxide, carbon monoxide and suspended particulate matter, Uruguay Round liberalization result in higher emissions in the developing countries (with the exception of sulphur dioxide increase in Japan for 1990 scenario) and lower emissions in the developed world for both the 1990 and 2000 scenarios. Nitrogen oxides emissions increase in all regions under both scenarios (with the exception of USA emissions in 2000). The impact of Uruguay Round liberalization on the emission levels is significantly greater in the second scenario (2000) than in the first scenario (1990) due to increased growth rates.

Finally, Cole et al. (1998) present the monetary estimates of the damages/benefits resulting from emissions of pollutants as a result of Uruguay Round liberalization. They use the monetary impact estimates provided by Fankhauser (1995) and Pearce (1993); these estimates are based on the likely impact of a tonne of pollution in terms of the cost of mortality, the cost of resultant illness amongst the exposed population, etc. Table 6.2 summarizes the information contained in two separate tables showing the monetary cost/benefit for developed and developing countries in Cole et al. Monetary damage estimates are as large as 1.2% of a region's estimated gains from the Uruguay Round. In both scenarios, at the aggregate level, developed countries benefit from trade liberalization, while the developing countries lose in terms of the associated monetary estimate of the changes in air quality. Nitrogen dioxide and carbon dioxide are exceptions, in which case both the developed and the developing countries end up with environmental deterioration in both scenarios. A final interesting result is that in 2000 scenario, developing countries suffer largely in case of suspended particulate matter, and developed countries show a large improvement in the same pollutant.

4. Case Studies on Individual Countries

In this part, we will briefly mention the experiences of several developing countries regarding the environmental impact of trade liberalization. These summaries are taken from Jha et al. (1999). Their study presents an extensive analysis of interactions between trade and environment for 11 developing countries. The countries include Brazil, China, Colombia, Costa Rica, India, Malaysia, Philippines, Poland, Thailand, Turkey, and Zimbabwe. We briefly summarize the experiences of countries with significant changes in environmental conditions that can be attributed to trade liberalization policies.

Brazil: In Brazil, main legs of the economic development policies since 1970s were import substitution, export orientation and openness to foreign direct investment (FDI). During the last two decades, exports increased rapidly, especially in the livestock, forestry and mining industries, which require high exploitation of natural resources. As compared to other developing countries, a large share of total exports of manufactured products comes from pollution-intensive industries. This share increased from 21.6 percent in 1970 to 44.4 in 1990. So, a general conclusion for Brazil is that expansions in manufacturing sectors and the sectors depending upon natural resources caused environmental deterioration after trade liberalization policies and economic development programs.

China: Impacts of trade liberalization in China are two-sided. As trade is more liberalized, the intrusion of environmentally friendly technologies into the economy increased. This has been complemented by the imports of more efficient power plants, mining machinery, metallurgical and casting equipment and light industrial machinery. These imports have been associated with the

Table 6.2. The Monetary Cost/Benefit Associated with the Change in Air Pollution Emissions for the Developing and Developed Regions (Millions of 1985 US$)

	Scenario 1 1990	Scenario 2 2000+0.5%
Nitrogen Dioxide		
Developed Countries	347	513
Developing Countries	206	1377
Global	553	1,890
Sulphur Dioxide		
Developed Countries	-227	-5350
Developing Countries	-13	662
Global	-240	-4,688
Carbon Monoxide		
Developed Countries	-13	-127
Developing Countries	2	17
Global	-11	-110
Suspended Particulate Matter		
Developed Countries	-814	-9427
Developing Countries	93	1110
Global	-721	-8,317
Carbon Dioxide		
Developed Countries	207	1451
Developing Countries	540	4836
Global	747	6,287

Source Cole et al. (1998)

need to use raw materials and energy more efficiently to accommodate the production boom resulting from trade liberalization. Similarly, increased demand for agricultural products paved the way for the application of methods such as crop rotation; these practices enhanced the ecological balance of the country. On the other hand, the detrimental effects of trade liberalization are more significant. Negative impacts of trade liberalization have been felt largely in the loss of biodiversity and the depletion of natural resources. Biodiversity is badly affected by the uncontrolled exports of wild animals and their products, such as frog legs, snake and weasel skins. Similarly, exports of wild plants resulted in different forms of environmental degradation. For example, exports of facai which grows on the border of deserts and has the vital role of sand fixation, has caused a desertification of over 200 km square. Increasing licorice root production has generated a reduction in reserves upto 40 percent of the 1983 level.

Colombia: The iron and steel industry was highly polluting and highly protected. After trade liberalization, the imports of steel have increased, and the domestic production has fallen. This had positive impact on the environment. On the other hand, with import liberalization, Colombia began to import US wastes for recycling, and the demand for domestic wastes has fallen; domestic wastes were dumped into landfills in Colombia. Export expansion of shrimps caused overexploitation. In case of coffee production, increasing prices in the world market forced the Colombian coffee producers to use high input systems, which have adverse environmental impacts. Finally, mass banana production induced by strong international demand has led to two environmental impacts: one is deforestation and the elimination of native species, and the other is application of highly toxic agrochemicals.

Malaysia: Malaysia's trade liberalization policies allowed her to increase the share of exports in GDP from 50 percent in 1970 to 81 percent in 1990; however, environmental problems began to occur with trade liberalization. One problem is the water pollution associated with the production of palm oil. Malaysian government introduced environmental regulations to control the changes in environmental conditions, and these regulations resulted in both improved water quality and export growth. In case of electronics industry, high rates of compliance attributed to multinational companies' installing emission controlling equipment, allowed control of potentially serious environmental problems. Nevertheless, the toxic wastes remain as the most significant problem. Accumulation of waste at other industrial sites, as a result of lacking treatment facilities, caused environmental problems.

Philippines: In the past restricted trade era, ban of export of logs from Philippines have resulted in wasteful investment in wood processing with high rates of wood waste, and import restriction of logs resulted in increased deforestation. In case of fisheries, as a result of tariff reductions, increase in the imports of fisheries have helped to reduce the pressure on domestic resources; however rising exports of crustaceans resulted in high production of crustaceans, and this had negative impact on wetlands. Table 6.3 shows the changes in the pollution intensity of Philippines exports over time. Clearly, we see that with trade liberalization, pollution intensity of exports declined over time.

Table 6.3. Philippines: Exports by Pollution intensity, 1975-1990 (percentages)

Pollution intensity	1975	1980	1985	1990
Highly polluting	51.07	31.30	38.36	21.84
Polluting	18.20	26.70	18.20	22.34
Less polluting	30.73	42.00	43.44	55.82
Total	100.00	100.00	100.00	100.00

Source: Intal et al. (1999).

5. Other Aspects

We finish the empirical exposition with some interesting works. Batra et al. (1998) touch an important but ignored area: energy intensity of international trade. Energy intensity is defined as the amount of energy used by an activity per dollar of its value-added. International trade necessitates the transportation of goods across vast land areas and the seas, which generate excessive amount of energy-related pollution. For example, air freight increased from 23% of commercial air traffic in 1970 to 25% in 1990 with the corresponding level of energy consumption rising from 4.2 million metric tons of aviation fuel to 11.2 million tons. Thus, trade related air freight fuel consumption almost tripled in just two decades. Batra et al. (1998) examine the impact of energy-intensity of trade on pollution. Although estimates of energy intensities are not available at the global level, the estimates for US show that trade is more energy intensive as compared to the overall economic activity and transportation. They consider a model where pollution is a function of energy consumption associated with domestic and global production and also with transportation of exports and imports. With pollution generating a direct utility cost to consumers, they demonstrate that trade with energy tariff or a global consumption tariff is superior to free trade.

Ferrantino (1997) presents some arguments in favor of a constructive interaction between trade liberalization and the environment. Protection in heavy industries (such as chemicals, iron and steel, pulp and paper) in developing countries generates environmental problems (as they can not use expensive cleaner technologies); these problems may be avoided by trade liberalization which would shift these industries to developed countries where cleaner technologies are in place, with an overall improvement in environmental quality in developing world and probably, globally. In a more specific case, trade liberalization may have positive impact on deforestation problem. Ferrantino (1997) states that the largest source of wood depletion in developing countries is probably wood gathering for household fuel and not the land clearing for farms. Higher incomes induced by trade liberalization will reduce household wood gathering by making other fuels affordable, and therefore, a decline in deforestation will be seen. Finally, freer trade in chemical and other agricultural inputs could increase the productivity of the existing land, and thus reduce the pressure to clear new land.

Chapter 7

GROWTH, TRADE AND ENVIRONMENT

In the previous chapter, we have explored different types of empirical studies related to the impact of trade liberalization on the environment. These studies ranged from single-country ones to multi-regional ones (including many countries, sometimes almost all important ones in the present context); they had different types of modelling (such as regular cross-country regression estimation, computable general equilibrium estimation). Some of these studies also incorporated economic development into the analysis; however, we deliberately delayed the role of growth in this literature to this chapter. In this chapter, we will explicitly analyze interactions between growth, trade and environment. In the literature, the standard approach is to model the *level of pollution* as a function of economic growth and trade liberalization (together with other relavant variables). Here, we will present an alternative approach and analyze the impact of trade and economic growth on the *environmental policy performance*. We will first start with a general introduction to the so-called environmental Kuznets curve (EKC) hypothesis; then, we will test it in a framework that is more consistent with the economic theory in the background. We will also incorporate the impact of trade liberalization on EKC hypothesis. In the second part of this chapter, we will investigate EKC hypothesis by using a recently developed econometric test known as threshold estimation. As EKC hypothesis is, in essence, about finding and testing the significance of a threshold, this technique is a perfect match for testing it. We start with a general introduction on EKC hypothesis.

S. Alpay (ed.), Trade and The Environment, 85–101.

1. Introduction to EKC Hypothesis

Understanding the impact of economic development and trade liberalization policies on the environmental quality is becoming increasingly important. The Stockholm Conference on Environment and Development in 1972 had been an important international meeting where concerns about global environment were outspoken, and the importance of formulating policies to overcome environmental problems started to be recognized. In 1980s and 1990s, with rapidly emerging concerns about global threats such as ozone-layer depletion and global warming, environmental issues made their way into public policy agenda in many developed countries. This is especially important nowadays as the environmental consequences of human activities exceeded certain limits, and can not be considered as negligible. On the other hand, economic development and trade liberalization are among the top priority policies in most of the developing countries. Thus, it is worth studying environmental consequences of economic development and more openness to trade.

In the literature, interactions between environmental quality and economic growth have been empirically modelled through emissions-income relationship by many authors. Empirical studies in this literature have been carried out in a specific format: different indicators of environmental degradation have been assumed to be an ad hoc polynomial function of income per capita, and then it has been tested whether there would be a decline in environmental degradation for income levels higher than a threshold. This search for an inverted-U type relationship between pollution and income, i.e. the environmental Kuznets curve hypothesis (EKC) has been at the center of discussion on the interaction between economic growth and the environment.

Grossman and Krueger (1991, 1993, 1995) have shown an inverted U-type relationship between per capita income and emissions of SO_2 and suspended particulates. EKC hypothesis has been tested by many others: Shafik and Bandyopadhyay (1992), Panayotou (1993b, 1997), Shafik (1994), Selden and Song (1994), Holtz-Eakin and Selden (1995), Suri and Chapman (1998), Kaufmann et al. (1998), Schmalensee et al. (1998), and Agras and Chapman (1999) can be seen among others. The results were mixed both in terms of an empirical support for EKC and the threshold level. The estimated threshold levels were substantially different across these studies ranging from \$2,894 (Panayotou, 1993b) to \$12,346 (Kaufmann et al. 1998). This large variation may be attributable to the differences in the source of data, inclusion of additional variables into the model, the use of emission or concentration of sulfur. Usually panel data from the Global Environmental Monitoring System's (GEMS) tracking of urban air quality in different cities in the developing and developed world have been used (Grossman and Krueger 1991, 1995, Panayotou 1997, Shafik 1994, Torras and Boyce 1996); OECD data was the next most commonly used data set (Cole et al. 1997, Selden and Song 1994). Technology level (Cole et al. 1997), locational dummies (Grossman and Krueger 1991, 1995, Shafik 1994), population density (Grossman and Krueger 1991, 1995, Panayotou 1997, Selden and Song 1994), GDP/area, imports/GDP, exports/GDP (Kaufmann et al. 1998, and Suri and Chapman 1998) were among the additional variables included in the models. Shafik and Bandyopadhyay (1992) have analyzed total and annual deforestation, where Cropper and Griffith (1994) have studied "rate" of deforestation. Selden and Song (1994) have looked at various air pollutants (suspended particulate matter (SPM), SO_2, NO_x and CO). Holtz-Eakin and Selden (1995) found that CO_2 emissions did not show an EKC pattern. Instead, CO_2 emissions monotonically increases with income (Selden and Song 1994). Hettige et al. (1999) have explored the income-environmental quality relation for industrial water pollution. They have shown that water pollution stabilizes with economic development, but have not detected an eventual decline.

Finally, the impact of trade liberalization on the environment has also been incorporated into the EKC framework in some studies. One can see Grosmann and Krueger (1991), (1993), Kaufmann et al. (1998), Taskin and Zaim (2001) and Agras and Chapman (1999).

2. Empirical Studies on EKC Hypothesis

One of the most extensive empirical studies on EKC is by Grosmann and Krueger (1995) examine the reduced form relationship between per capita income and various environmental indicators, including urban air pollution, the state of the oxygen regime in river basins, fecal contamination of river basins, and contamination of river basins by heavy metals. Their objective is to determine whether environmental quality deteriorates steadily with economic growth. Their findings have strong implications for the impact of trade liberalization on the environment as economic growth is generally associated with more openness (see, for example, Dollar 1992, Edwards 1992, and Harrison 1996; Harrison's study incorporating vector auto-regression suggests that the causality between growth and openness runs in both directions). They estimate the following reduced-form equation:

$$Y_{it} = G_{it}\beta_1 + G_{it}^2\beta_2 + G_{it}^3\beta_3 + \bar{G}_{it}\beta_4 + \bar{G}_{it}^2\beta_5 + \bar{G}_{it}^3\beta_6 + X_{it}\beta_7 + \epsilon_{it}$$

where Y_{it} is a measure of water or air pollution in station i in year t, G_{it} is GDP per capita in year t in the country in which station i is located, $\bar{G}_{it}$ is the average

GDP per capita over the prior three years, X_{it} is a vector of other covariates (like temperature, population density), and ϵ_{it} is an error term.

They used the panel data in the Global Environmental Monitoring System's (GEMS) tracking of urban air quality in different cities in the developing and developed world, and the panel data from the GEMS monitoring of water quality in river basins around the globe. The panel data from the Global Environmental Monitoring System's (GEMS) tracking of urban air quality in different cities in the developing and developed world runs from 1977 upto 1991 with a total number of 1478 observations. The data includes 47 cities in 28 countries in 1977, 52 cities in 32 countries in 1982, and 27 cities in 14 countries in 1988; overall, 42 countries are represented in the sample. Water quality data, especially on rivers, includes a large number of stations in 58 different countries. GEMS data is an extensive data and it includes many different pollutants such as sulfur dioxide emissions, smoke, suspended particulates etc. and related to water pollution it contains observations on dissolved oxygen demand, coliform, etc. Estimation has been done by using GLS method to account for any other characteristics that are not included in their list of regressors.

They find no evidence that economic growth harms the natural habitat steadily. Rather, they determine an inverted-U type relationship between growth and environmental degradation for most of the environmental indicators they have used: economic growth brings an initial phase of deterioration which is followed by a subsequent phase of improvement. The turning point of this inverted-U type relationship is on average around an income of $8000 per capita. More specifically, for a country with a per capita income of $10000, the hypothesis that further growth will generate environmental degradation can be rejected at the 5 percent significance level for many of the pollution measures they have used. The important question that comes to mind is what linkages between income and environment exist to explain the results obtained. Their answer comes from a review of the available evidence on instances of pollution abatement by OECD (1991) which suggests that as countries experience greater prosperity, their citizens' demand for higher environmental quality rises. Nevertheless, as Grosmann and Krueger (1995) mention in their paper, it is unclear why the estimated relationship between pollution and income exists. A more consistent but also more complex analysis, would be to model the structural equations relating environmental regulations, technology and industrial composition to GDP, and then to link the level of pollution to the regulations, technology and industrial composition. Such structural approaches are undertaken by Dean (1998) and Antweiler et al. (2001), summarized above in chapter 6. Grosmann and Krueger (1995) also state that inverted-U relationship needs to be taken very carefully since it depends on the type of the polluting particle; increase, for example, in solid waste, and ground water contamination in developed countries continues with economic growth (see, for example, World Bank 1992). In support of this, Ekins (1997) uses an aggregate indicator of environmental impact developed by OECD which incorporates 10 different environmental measures, in the analysis of EKC hypothesis. He finds no evidence of inverted-U type relationship between this indicator and the income.

In an earlier paper, Grossman and Krueger (1991) also included variables representing openness to trade in their model. The approach was very similar to the one described above, and they concluded that SO_2 levels were significantly lower in cities of the countries that conduct a great deal of trade. In an analogous EKC study, Shafik and Bandyopadhyay (1992) test the hypothesis that the more open a country's trade regime the cleaner will be her production processes. Their results are mixed, and at best, they find weak evidence that more open economies pollute less. Other EKC studies related to our subject matter will be summarized in the next section.

3. An Alternative Test of EKC

All these studies try to establish a direct linkage between income and pollution and/or between trade and pollution. They seem to overlook the more basic and fundamental interaction among these variables which is the impact of income growth and trade liberalization on environmental awareness and policy making. Theoretically, if one considers environmental quality as a normal good, one would expect that demand for better environment and therefore public pressure for stricter environmental regulations will rise with increases in per capita income. We examine this important linkage between environmental awareness, regulations and income empirically. In particular we focus on four questions: (1) whether income is a significant determinant of environmental awareness, environmental policy making and performance, (2) whether this interaction shows different characteristics at different stages of the economic development, (3) whether environmental Kuznets curve hypothesis (EKC) can be supported by considering the impact of income on the environmental policy performance (which is a direct test of the economic theory behind EKC), and (4) whether trade liberalization lead to better environmental policy performance or not.

In interpreting the direct relationship between income and environmental degradation, two key explanations are provided in the EKC literature. Firstly, with higher income, the structural change in the production sector is expected to lead to a decreased share of pollution intensive sectors; secondly, increases in income will raise environmental awareness, and so environmental regulations will be more stringent. However, these linkages have not been formally tested to support EKC. It is obvious that income is not a direct determinant of environmental quality. Higher income may lead to better environmental awareness and this may be reflected in environmental regulations. Only through effective implementation of these regulations, improvement in environmental quality can be achieved. We may attribute the lack of studies along this line to the difficulty in obtaining a suitable index for environmental awareness and policy making. Nevertheless, there has been several attempts for such a measure. Walter and Ugolow (1979) and recently van Beers and van den Bergh (1997) have developed indices for the stringency of the environmental policies. Dasgupta et al. (1995) have taken up a more comprehensive research, and developed an index for environmental policy performance (EPP). The EPP index is constructed based on the following five components: environmental awareness, scope of policies adopted, scope of legislation enacted, control mechanisms in place, degree of success in implementation. For each component a number of questions have been used to derive the score for that component. Then the scores for these five components have been added up to get the overall EPP index. The answers to questions have been derived from the country reports prepared for the United Nations Conference on Environment and Development (UNCED, 1992). Since the country reports were similar in form as well as in coverage (UN imposed a standard reporting format), the index could be derived with consistency across countries. More specific information will be provided below.

Dasgupta et al. (1995) have also studied interactions between income and environmental policy performance along with the development of EPP index, and have determined a positive relationship; however, the set of countries they used did not show adequate variation in income: there were only six countries with income levels larger than 5,000 US dollars; 4 countries with income between 2,000 and 5,000 US dollars, and remaining 21 countries had income levels below 2,000 US dollars. They have not analyzed income—environmental performance interaction across different income levels either as that was not very related to the main theme of their paper; thus, their study may not be easily used to replace the standard EKC studies in the literature. Eliste and Fredriksson (1998) have extended the set of countries, and removed the low-income-country bias in the data set, but have been limited due to its focus only on the agricultural sector. Nevertheless, given these two extensive works on the measurement of environmental performance, it is not very difficult to overcome these problems and to perform an alternative test of the EKC hypothesis as indicated by the basic economic theory.

While making use of these studies to test EKC, we will also include some trade liberalization variables in our model. Eliste and Fedriksson (1998) considered the impact of more openness on the environmental policy performance in agricultural sector in order to disclose whether countries relax domestic standards for environmental quality to increase (or maintain) "competitiveness" (a "race to the bottom"), or even discourage the enactment of environmental policies altogether (a "regulatory chill"). They have used the methodology of Dasgupta et al. (1995) to obtain environmental policy performance index (related to agricultural sector only) as their dependent variable, and three different direct measures of openness of trade regime (trade intensity, Sachs and Warner index and Fraser openness index) as the explanatory variables along with GDP per capita, farmers' lobbying pressure, fertilizer use, country interaction variables, freedom of information and level of democracy dummies. It has been shown that countries with more open trade regimes have more stringent regulations and better environmental performance (however, trade intensity indicator had no significant impact).

Eliste and Fredriksson (1998) have carried out their analysis on the environmental policy performance only in agricultural sector. As they point out, the agricultural sector may be a special case because of food safety standards and other sanitary measures, and because it is a resource based (i.e., with immobile capital) sector where lower environmental regulations do not induce great capital movement. Thus, it would be a worthwhile attempt to extend their study in a setting where the impact of more openness on overall environmental policy performance can be demonstrated. We will include this extension in our study.

3.1 Data and Model

The key variable of this study, the environmental policy performance (EPP) index, has been derived from the studies of Dasgupta et al. (1995) and Eliste and Fredriksson (1998). The data source of both studies was the reports presented by a large number of countries to the United Nations Conference on Environment and Development (UNCED) 1992. Dasgupta et al. (1995) developed an overall environmental policy performance index for 32 countries and Eliste and Fredriksson (1998) computed environmental policy performance index for 60 countries but only for the agriculture sector.

The reports were prepared according to a standard reporting format imposed by UN, and thus they were comparable across countries. The country reports provide sector specific information about the state of the environment and natural resource utilization by the agriculture, industry, energy, transport and the urban sectors. Information on existing environmental policies, legislation, control mechanisms and enforcement have been provided. Using this information,

Dasgupta et al. assessed the state of: "(i) environmental awareness; (ii) scope of policies adopted; (iii) scope of legislation enacted; (iv) control mechanisms in place; and (v) the degree of success in implementation" through a multidimensional survey. Twenty-five survey questions are included, and one of 0, 1, 2 values are chosen for each question, where these values indicate low, medium and high performance, respectively. The results are summarized in a matrix form consisting of the four environmental aspects (Air, Water, Land and Living Resources) in one dimension and five activity sectors (Agriculture, Industry, Energy, Transport and the Urban sector) in the other. Then by summing up all of the 500 entries for each country, overall index is obtained.

Alpay et al. (2001) extended the data set in Dasgupta et al. (1995) from 32 countries to 60 countries by using Eliste and Fredriksson (1998) as the set of countries included in the former was mostly consisting of low-income countries. Since we were unable to reach the actual reports used by Dasgupta et al. (1995) and Eliste and Fredriksson (1998), we made use of these studies in the following manner to get an overall environmental performance index for 60 countries: We regressed overall environmental performance index of Dasgupta et.al. (1995) to the agricultural one for the countries common in both Dasgupta et al. (1995) and Eliste and Fredriksson (1998). Then, using the estimated regression equation we have obtained the overall environmental performance index for the remaining countries in Eliste and Fredriksson (1998).

Our data sources for the remaining variables are as follows: GDP per capita expressed in 1985 international prices was obtained from Penn World Tables (PWT). To get rid of temporary fluctuations in the income, we took three-year averages (1990, 1991 and 1992). Other data obtained from this source include population, and trade intensity. The urbanisation data was obtained from "World Resources 1992-93" published by World Resources Institute. For missing countries in World Resources, we resorted to World Development Indicators 2000, published by the World Bank.

Various indicators of openness to international markets have been suggested in the literature. The most widely used one is the ratio of the sum of exports and imports to GDP (called as trade intensity–TI). We also included other indicators of openness: black market premium in foreign exchange markets (BMP), average level of tariffs on imports (TAR), and coverage of quotas on total imports (QUO) (measured by the percent of imports affected by a quota).

Our proposed model for the determinants of environmental policy performance is as follows:

$$EPP = f(GDP, URB, TI, BMP, TAR, QUO) \tag{7.1}$$

where EPP is environmental policy performance index, GDP is the GDP per capita, URB is the urbanization rate, TI is the trade intensity (defined as the ratio of sum of exports and imports to GDP), BMP is the black market premium in foreign exchange markets (can also be viewed as an indicator of financial liberalization), TAR is the tariff coverage for imports, QUO is the quota coverage. BMP is a dummy variable in our model so that the value 1 indicates there is no positive or negative premium in foreign exchange markets (indicating financial liberalization), and 0 indicates lack of financial liberalization.

An important objective of our analysis is to identify EPP-economic development relationship at different income groups to be able to test the EKC hypothesis. For specifying the relevant categories of income, we have made use of the scatter diagram of GDP per capita and EPP. We identified a reasonable grouping based on both the scatter diagram and the commonly-agreed income grouping of countries as follows (note that approximately even distribution of points to each category is also targetted): Low-income between \$0-\$2000, Middle-income between \$2000-\$8000, and High-income above \$8000.

To see whether the intercept and slope of the EPP-income relationship is significantly different in these groups, we include two intercept dummies, LI and MI and two slope variables GDP*LI and GDP*MI. LI is 1 if the country is a low-income one and 0 otherwise. MI is 1 if the country is a middle-income country and 0 otherwise. In our initial regressions where we included all the variables listed at the beginning of the section as explanatory variables, the variables URB, TAR, QUO were insignificant, and thus, they are disregarded from the regression equation. Our modified model in line with the discussion above is as follows:

$$\begin{aligned} EPP &= \beta_1 * GDP + \beta_2 * TI + \beta_3 * BMP + \beta_4 * LI + \beta_5 * MI + \\ &\quad \beta_6 * GDP * LI + \beta_7 * GDP * MI + \varepsilon \end{aligned} \tag{7.2}$$

where ε is the error term. We now summarize the estimation results.

3.2 Estimation Results for EPP

Results of the OLS regression of equation (7.2) are as follows:

$$\begin{aligned} EPP &= 0.02 * GDP + 0.67 * TI + 563 * BMP + 270 * LI + 278 * MI \\ &\quad + 0.08 * GDP * LI + 0.03 * GDP * MI + \varepsilon \end{aligned} \tag{7.3}$$

Intercept term has been dropped because it had perfect correlation with the dummies (BMP, LI, and MI) included in the model. Number of observations has to be reduced to 51 due to lack of data on BMP variable for some countries; we have also removed outliers from the data set after initial estimations. All the estimated coefficients are significant at 5 percent level. We find that per capita income has a strong and positive impact on EPP. This result verifies the findings of Dasgupta et al. (1995) and of Eliste and Fredriksson (1998). As people get richer, they demand better environmental quality, and so stricter environmental regulations. Secondly, the trade intensity (TI) variable and financial liberalization dummy (BMP) have significant positive effect on EPP. This implies that more openness leads to better environmental policy performance. Finally, we see that the designated income groups show *significantly different* characteristics in terms of EPP–income relationship. To relate our findings to the EKC hypothesis, we focus on the impact of income on EPP for the low, middle and high-income groups separately:

Low-income: 0.10 (=Coefficient of GDP + Coefficient of GDP*LI)

Middle-income: 0.05 (=Coefficient of GDP + Coefficient of GDP*MI)

High-income: 0.02 (=Coefficient of GDP)

Looking at these coefficients, the sensitivity of EPP to income among the middle and especially low- income group is very high. This implies that there is a significant potential for better environmental policy performance in the low and middle-income groups as compared to high-income group. This differential interaction between income and EPP at different stages of economic development also helps us to interpret our results along the EKC hypothesis. We leave this discussion to our concluding remarks section.

Our second result that more openness leads to better environmental policy performance is similar to those of Antweiler et al. (2001) and Eliste and Fredriksson (1998). This positive impact can be as a result of the shift of production processes of the low-income countries to cleaner ones in order to meet higher environmental standards of the importer high-income countries; this way enhancement of policy performance becomes easier for these countries. Furthermore, international environmental agreements may have a real positive effect on environmental policy performance. Especially, practices like blocking exports from non-ratifying exporters seem to be effective. Finally, more openness may be leading to increased awareness on environmental matters due to higher interactions among countries.

4. A Threshold Test of EKC Hypothesis

Our goal in this section is two-fold. Firstly, studies on EKC hypothesis are criticised due to their choice of the quadratic or cubic specification of pollution with respect to income per capita. It is unclear why the specific reduced-form equation employed in their estimations exists. Our first contribution is to overcome this problem by employing the threshold estimation method developed by Hansen (2000). This new method models threshold directly and it is easier to interpret economic relationships compared to a polynomial model. Threshold model is a simple, parsimonious non-linear model. It is easy to understand, and compared to other non-linear models its application is simple. This allows for non-linearities in conditional expectation function. It is also a sub-case of more complicated Markov Switching models. Schmalensee et al. (1998) study uses spline (piecewise linear) estimation, and thus avoids the quadratic or cubic specification; however, their model does not provide any theoretical reasoning for the division of the data into income segments except that each segment contains equal number of observations. Moreover, there is no statistical test on the *existence* and *significance* of the threshold. Their approach takes threshold level as pre-determined (the income segments are chosen by the authors) instead of finding it from the *data*. We will address these critical issues in this section.

Secondly, as indicated by Diwan and Shafik (1992), and Rothman (1998), trade makes the separation (disassociation) of domestic consumption and production decisions possible; thus, a country can secure sustainable development (in terms of its environmental quality) at the expense of unsustainable development in other countries. We believe that empirical investigation of this is quite important, and we will present an explicit test of the hypothesis that there is an inverted-U type relationship between pollution and trade liberalization; that is to say, we will test the existence of an environmental Kuznets curve hypothesis between pollution and more openness. If an EKC type behavior exists, then one needs to look deeper into the source of this finding, and the possible shift of pollution-intensive industries across countries through trade will be among the alternative explanations. Our approach is critically different from earlier studies incorporating openness-to-trade variables into the EKC studies. Earlier studies use openness-to-trade variables only as independent variables and not as the *threshold* variable. We will determine if there is any significant threshold with respect to openness-to-trade, and if it exists, how pollution and openness relate to each other before and after the threshold.

4.1 Introduction to Threshold Estimation

We consider the following threshold regression model by Hansen (2000):

$$Y_i = \beta_1 + \beta_{21} X_i 1_{(Xi \leq \lambda)} + \beta_{22} X_i 1_{(Xi > \lambda)} + \beta_3 Z_i + \epsilon_{it}$$

where $X_i$is the threshold variable, the variable that causes the regime shift,Z_i represents the rest of the independent variables, λ is the threshold level, and $1_{(.)}$ is the indicator function ($1_{(.)}$ is 1 if the condition in (.) is satisfied, otherwise it is zero).

The threshold test is performed by testing the equality of coefficients β_{21} and β_{22}. The next step is the estimation of these coefficients if a threshold is found to be present in data. Estimation is done by simple least squares (LS) regression.

4.2 Model and Estimation

After introducing the new econometric technique that we will employ in our analysis, we will now introduce its application in the test of EKC hypothesis. Our estimated model is almost same as the one in Grossman and Krueger (1995) described above with the exception of the removal of polynomial specification of income terms. Incorporating the threshold technique introduced above, we estimate:

$$Y_i = D_i\beta_1 + G_i 1_{(Gi \leq \lambda)}\beta_{21} + G_i 1_{(Gi > \lambda)}\beta_{22} + CO_i\beta_3 + R_i\beta_4 + CC_i\beta_5 + I_i\beta_6 + \epsilon_i \quad (7.4)$$

Y_i is a measure of water or air pollution in station i , D_i is population density, G_i is GDP per capita in the country in which station i is located; CO_i is coastal dummy; R_i, CC_i and I_i are other location dummies indicating residential, center of city, and industry, respectively; ϵ_i is the error term. Location dummies are removed in case of pollutants where they are not relevant (such as biological oxygen demand and fecal contamination). All these variables are included in the Grosmann and Krueger (1995) model above under X_{it}. In the first stage of our estimation, we will try to identify whether a threshold exists with respect to income variable. If a significant threshold exists, then the coefficients in the above model are estimated both before and after the threshold GDP level.

Given the requirements of the threshold model we are employing, we can not pool the cross section and time series data as in Grosmann and Krueger, and thus, we separate the data across years, and repeat the estimation for each year.

Because of the lack of enough observations, we could not obtain threshold estimates for some of the years and for some of the pollutants used by Grosmann and Krueger (1995) (for a reliable threshold estimation, one needs at least 100 observations). We have used the same GEMS data as in Grosmann and Krueger (1995). The panel data from the Global Environmental Monitoring System's (GEMS) tracking of urban air quality in different cities in the developing and developed world includes data collected from 42 different countries. Water quality data, especially on rivers, includes a large number of stations in 58 different countries.

Table 7.1. Threshold Estimation for Sulfur Dioxide Emissions

Year	Beta Before Threshold, β_{21}	Beta After Threshold, β_{22}	Threshold	Number of Observations
1977	45.58 (1.233)*	2.676 (1.339)*	2883[a]	123
1978	11.248 (1.902)	1.461 (0.349)	4422[a]	126
1979	3.331 (1.180)	-0.084 (0.560)	10423 [b]	127
1982	10.261 (2.154)	1.20 (0.483)	4591[a]	142
1984	15.675 (3.927)	0.390 (0.551)	3526 [b]	145

Source: Alpay and Caner (2002).
* Standard errors in parenthesis.
[a] Significant at 0.001.
[b] Significant at 0.01.

As clearly seen in Table 7.1, a significant threshold exists for all years used in the estimation; however, there is hardly any evidence for EKC-type behavior. With the exception of 1979, the effect of GDP on pollution level is positive both before and after the threshold level. Although, the positive impact shows significant decline, there is no reversal in the impact of GDP on pollution level.

We have also applied threshold estimation for some other pollutants for which Grossmann and Krueger (1995) identified EKC type behavior. In case of biological oxygen demand (BOD), no significant threshold exists for the years 1982—1985 (for other years, number of observations were not adequate). Another indicator of water quality is dissolved oxygen. We could test for threshold for three different years. In 1981 and 1982, there exists a significant threshold; however, only in 1982 we get barely an EKC-type behavior (in 1981, coefficients of income term before and after the threshold were both positive). For 1985, no threshold was present. Finally, we also tested the existence of EKC hypothesis for the fecal coliform. For 1981, 1983, and 1985, we were not able to get a significant threshold. For 1980, we identified a significant threshold, but no EKC-type behavior. Only for 1982, there was a significant threshold and EKC-type behavior.

4.3 EKC Hypothesis Between Trade and Pollution

The impact of trade linkages among countries has been studied only in a few studies in the context of EKC. As indicated by Rothman (1998), international trade provides the means through which domestic consumption and production can become disassociated; thus, it is perfectly possible for a country to reduce its environmental degradation by shifting all or some of its dirty industries to other countries (Diwan and Shafik 1992). Therefore, the role of trade in the analysis of EKC-type behavior between economic growth and environmental degradation should not be neglected. The downward movement in the environmental degradation, if exists, may be due to increasing volume of trade, and not due to the increasing income. Moreover, this way of reducing environmental degradation will obviously not be available to the latest-developing countries (Ekins 1997).

We would like to apply the threshold estimation technique to this issue as well. Towards that end, we have obtained openness-to-trade indices for the countries included in Grossmann and Krueger (1995) study. This index is defined as the ratio of sum of exports and imports to GDP; it is also known as trade intensity, and used in earlier studies as well (see, for example Antweiler et al. 2001). Then, we have estimated the following model with the threshold technique: (for each of the years analysed)

$$\begin{aligned} Y_i = \; & D_i\beta_1 + G_i\beta_2 + CO_i\beta_3 + R_i\beta_4 + CC_i\beta_5 + I_i\beta_6 + \\ & OP_i 1_{(OP_i \leq \lambda)}\beta_{71} + OP_i 1_{(OP_i > \lambda)}\beta_{72} + \epsilon_i \end{aligned} \quad (7.5)$$

where Y_i, D_i, G_i, CO_i, R_i, CC_i , and I_i are as defined above; OP_i is the openness index in the country in which monitoring station i is located. Note that there is an important distinction between this equation, and the previous one other than the inclusion of openness index: the threshold variable is now changed to openness index from GDP per capita. If we can detect a significant reversal in the sign of openness index then one can talk about an EKC hypothesis in the level of openness to trade as opposed to per capita income. We have tested for the threshold in openness only for sulfur dioxide emissions and dissolved oxygen due to data limitations. In the case of dissolved oxygen, we have identified a significant threshold for openness index for 1979 and 1980; the threshold for 1981 and 1982 was not statistically significant; however, in all cases, no evidence exists for the EKC type behavior between environmental quality and openness to international markets. Estimation results for sulfur emissions are shown below:

Table 7.2. Estimation Results for Threshold in Openness to Trade

Year	Before Threshold, β_{71}	After Threshold, β_{72}	Threshold Level	Number of Observations
1977	0.01 (0.006)*	-0.077 (0.011)*	28.42^b	118
1981	0.137 (0.023)	-0.0015 (0.005)	24.21^a	132
1983	0.036 (0.012)	-0.005 (0.006)	10.76^c	110
1984	0.561 (0.111)	-0.0007 (0.007)	19.5^b	112
1985	0.281 (0.070)	-0.007 (0.006)	16.43^b	97

Source: Alpay and Caner (2002).
* Standard errors in parenthesis.
[a] Significant at 0.001.
[b] Significant at 0.01.
[c] Significant at 0.05.

As seen in Table 7.2, there is somewhat strong evidence for the existence of a threshold in openness index. We also have a weak support for EKC hypothesis in openness, that is to say, up to a certain openness level the impact of higher trade liberalization on the environment is negative, and after that level, higher liberalization results in lower environmental degradation; however, note that although coefficients after the threshold are always negative, they are not significant (with the exception of 1977) due to relatively high standard errors.

5. Summary on EKC Findings

In this chapter, we have presented empirical analysis of interactions between economic growth and environmental degradation. Earlier empirical studies in this literature bypassed the implications of economic theory, and concentrated on income—pollution relation directly. From these empirical studies, environmental Kuznets curve hypothesis has emerged indicating an inverted U-type relationship between income and pollution. In the first part of this chapter, we provide an empirical test of the theory as it stands, and focus on the relationship between income and environmental policy performance. We argue that this investigation is more critical and deserves more attention than income-emissions relation as the change in emission concentrations is a secondary effect that occurs *after* income shows its effect on policy changes. We also analyze the implications of trade liberalization for the environmental policy performance (EPP) by using various openness indicators.

Firstly, our findings related to income-EPP relationship indicate that for low-income countries there is a significant potential to improve their policies on pollution control as their respective economies grow. Secondly, our empirical results show that per capita income has a very strong and positive relation with environmental policy performance (EPP). Moreover, our classification of countries with respect to their income turned out to be useful as the income-EPP relation was found to have different characteristics across income groups. Marginal contribution of income to the level of environmental policy stringency is shown to be higher in low income countries as compared to both middle income and high income countries. Noting that the level of EPP is higher in high-income countries than in middle and low-income ones, this may be used as an evidence for EKC hypothesis. The decline in marginal contribution of income to EPP with rising income indicates the possibility that higher income countries have already taken enough precautions for a better environment so that there is limited room for additional policy improvement that may be generated with even higher income. Additionally, given that environmental quality is a normal good, if the pollution levels were not declining at higher income levels compared to lower income levels, then there would be room for new policy interventions, which would be contradicting our findings. The changing nature of the relationship between income and environmental policy performance (EPP) will most likely imply a changing interaction between emissions and income at different income levels. Nevertheless, one needs to perform an explicit test between EPP and pollution levels in order to link these findings to the EKC hypothesis in a profound manner.

We have also presented an empirical test of the hypothesis that trade liberalization induces a "race to the bottom" in the political determination of environmental regulations. This has been tested before by Eliste and Fredriksson (1998), but only for the agriculture sector. Given the constraints imposed by the choice of agricultural sector explained in earlier sections, extension of their study in a setting where the impact of more openness on overall environmental policy stringency has been analyzed, is valuable. We have shown that the impact of more openness on environmental policy stringency was positive and significant for both trade intensity and the financial liberalization indicator; therefore, "race-to-the-bottom" hypothesis cannot be supported empirically in this more general setting either. We can interpret this positive association as the effect of a greater economic surplus to use for environmental protection, reputational effects, increased technology transfers, and a greater exchange of ideas about environmental regulations as stated in Eliste and Fredriksson (1998).

In the second part of the chapter, we have addressed a modelling criticism related to empirical studies on EKC hypothesis: the second or third order polynomial specification of the pollution with respect to income without any economic theory in the background. We remove this modelling problem by employing a recently developed econometric technique by Hansen (2000) for the estimation of thresholds in the data. The threshold econometric model, a sub-case of more complicated Markov Switching models, is simple to apply, and easy to understand. It is essentially running two separate linear regressions before and after the threshold, if a threshold is found to exist in the data.

As a general conclusion, our estimation results do not support the existence of EKC type behavior between economic development and environmental quality. Although we have used the same data set as in Grosmann and Krueger (1995), our findings are not supportive of theirs: a threshold exists in most cases, but the impact of income on the level of emissions does not show a reversal in sign; at best, the positive impact becomes smaller. Increases in income reduce the load on the environment, but they do not lead to improvement in environmental quality.

An important issue in this literature is the role of trade liberalization policies in the framework of EKC hypothesis. International trade makes the separation of consumption and production activities possible, and thus, pollution intensive industries can move across countries. Then, one may talk about two possible sources for an improvement in environmental conditions: higher openness to trade and/or higher income. Our second contribution is the explicit test of the hypothesis that higher openness leads to improved environmental conditions, in other words, a test of EKC behavior between openness to trade and environmental quality. Our findings present some weak support for this hypothesis for sulfur emissions; however, for dissolved oxygen indicator, no evidence exists. In case of sulfur emissions, where there is a significant threshold, all the countries that fall into the upper threshold region are developed countries (with per capita income levels at least 7000 dollars); in the lower threshold region, there are a few developed countries, but most of them are developing ones. Thus, we have some reason to think that our estimations present a weak support for the hypothesis that sustainable development (as regards to environmental quality) in developed countries is being achieved at the expense of unsustainable development in developing ones. Such a trend, if exists, can not be sustained globally. We suggest the investigation of this interaction between openness and environmental quality more extensively by using better and more recent data sets. Future work in this literature should try to address whether improvement in environmental conditions is due to economic development (higher income), or openness to international markets or both.

Chapter 8

IS TRADE GOOD FOR THE ENVIRONMENT? GENERAL CONCLUSIONS

In the preceeding chapters, we have presented an introduction to trade and the environment literature from the perspective of the impact of trade liberalization on the environment. Many theoretical and empirical studies that investigate the environmental consequences of freer trade with different modelling approaches provide evidences for the two possible outcomes: (1) trade liberalization generates positive environmental side effects, and (2) trade liberalization increases the environmental degradation. Thus, a universal conclusion related to the environmental impact of trade liberalization will be dubious.

Rather, interactions between trade liberalization and environmental quality depend on country-specific factors, and the existing institutional structures, especially related to the environment. We may list following set of factors as the key determinants of this interaction: (i) whether the country in consideration is a developed or developing one (demand for environmental quality is a positive function of income); (ii) the comparative advantage of the country implementing liberalization policies; (iii) resource intensity of the traded product and the property-rights structure associated with the use of the resource; (iv) existence and implementation of environmental policies during trade liberalization; (v) whether environmental policies are affected by strategic trade considerations; (vi) whether environmental externality is consumption-based or production-based; and (vii) type of the pollution, transnational or local. In fact, all these issues can be considered as the points of departure among the studies in this literature, and naturally, they may be seen as the main constituents of the contrasting results summarized in the preceeding chapters.

Due to differences in the environmental policy stringency levels between developed and developing countries, most developing countries will have com-

S. Alpay (ed.), Trade and The Environment, 103-108.

parative advantage in pollution intensive goods; thus, production of these goods may easily shift to the developing countries following trade liberalization policies, and consequently, will result in increased pollution both locally and globally. It has been shown strongly that the differences in pollution policies would generate incentives to trade, and one should seriously consider further environmental deterioration in countries with lax regulations following trade liberalization. Therefore, from an environmental perspective, it is important that trade liberalization policies should be accompanied by some kind of "harmonization" of environmental policies across countries; however, given the disagreements between developing and developed countries on the process towards "harmonization", this seems to be not very plausible at this point. We should also note that "harmonization" of policies should not be understood as one-to-one matching of policies across countries. Differences in country characteristics such as assimilative capacity, have to be taken into consideration.

Removal of tariffs in highly protected polluting industries in developing countries or removal of export subsidies in those sectors will open ways for improved environmental quality; however, this kind of trade liberalization policies can not be easily implemented by the governments of developing countries as it implies loss of competitiveness, and so, of jobs in these industries in developing countries. To make this shift possible, some kind of trade adjustment programs between developed and developing world seems to be required.

Expected increases in income associated with freer trade will generate demand for environmental quality and thus, will put downward pressure on emission levels. Nevertheless, this seemingly automatic transition is subject to some criticism. If the pollution in high income countries is being reduced by shifting polluting industries to lower income countries, this kind of pollution reduction will not be available to the latest-developing countries as there will be no countries coming up behind them to which this relocation can be directed . Thus, installation of better technologies and shift to cleaner production processes should be given priority; again, the way how this will be done (in terms of funding) will be a hot debate topic between developed and developing countries. Existence of studies which show that a negotiation linking trade and environment policies may be better than a trade-only negotiation for developing countries, but compensation for environmental restraint would be even better for them, makes this issue more interesting.

Additionally, in the presence of a negative production externality (in particular, industrial pollution affecting the outcome of an environmentally sensitive industry, such as fishing), international trade can play a useful role in spatially separating these incompatible industries across countries with complete specialization.

On the other hand, in the absence of environmental policies, increased economic activity associated with trade liberalization will obviously put upward pressure on emission levels in countries with comparative advantage in dirty industries. Changes in the composition of production stemming from freer trade may relocate dirty industries to developing countries from developed ones; although an improvement is observed in developed countries after trade, on the global level the emissions may increase. Production shift resulting from trade is of no use for the environment if it is not associated with changes in the consumption behavior away from pollution intensive goods. The pressures associated with the competition of domestic firms with their foreign rivals may lead to relaxation of the stringent environmental policies, or may delay the enactment of stringent environmental policies if they do not exist at the time of trade liberalization.

Finally, the balance in the use of natural resources will be destroyed if these are opened to international markets without any reservations, and in particular, in the absence of clearly defined property rights, which is the case in many developing countries. Linkages between increases in the volume of trade and deforestation associated with the production of cash-crops in some developing countries are well known. Given the importance of these resources for the global environment, protective environmental policies need to be enacted with freer trade in these products. Practical evidence shows that cooperation between developing and developed countries on this issue has to be established as well.

These are some of the main channels, suggested in the literature, through which trade liberalization will have impact on the environment. Luckily, some countries may observe improved environmental conditions with higher liberalization without taking any precautions; however, a wiser approach will be to complement trade policies with environmental ones; most of the studies we surveyed require an active environmental policy for a better environmental quality after trade liberalization. Since developing countries pay normally more attention to economic growth, and since environmental policies are seen as an obstacle for growth, active cooperation between developed and developing countries is required. This cooperation will have to be arranged by considering the interactions between trade and environment mentioned above, and by making sure that when designing their economic development and trade policies, developing countries will not take the environmentally misguided path of developed countries during their industrialization process. Among other things, this seems to require financial assistance to developing countries from developed ones as indicated by developing countries in many debates on this issue, and also as seen practically in some of the successful international environmental agreements (for instance, Montreal Protocol).

Another interesting conclusion of our exposition, at least for the short and medium terms, is that the impact of trade liberalization on environment casts serious doubt on the well-known gains from trade argument. Not only can trade liberalization generate environmental deterioration, but also an overall welfare decline can be observed (given that environmental quality is included in the welfare expression). Unconditional superiority of trade liberalization should be taken with precaution; the policies toward more open economy will serve the desired objectives better if they are accompanied by environmental policies.

A collection of papers that are constructively critical of the gains from trade argument from an environmental perspective has been edited by Ekins et al. (1994). Their most significant arguments include: (i) for economic growth to benefit the environment, it is not enough to generate additional resources, but those resources must be targeted towards environmental quality; a common observation in outward-oriented economies is that this is not being realized; (ii) even if some of these additional resources are directed to environmental protection, nothing can be done about irreversible damages of economic growth to the environment, and (iii) increases in the volume of transportation generated by trade liberalization will contribute substantially to energy-related environmental damage.

We think that further research in the following areas will be very useful. Firstly, to improve our understanding of the interactions between trade liberalization and environmental quality, we need more empirical studies at the individual country levels. This way, the country specific factors, such as development level, population, income distribution, that affect the links between trade and environment will be taken into consideration. With that information at hand, the policies on trade and the environment can be made more constructive and supportive of each other.

Concerns of developed and developing countries on trade and environment issues are asymmetric, i.e., better trading conditions are more important for developing countries whereas higher environmental quality for developed world. There exists somehow a complementarity between trade and environmental policies stemming from the asymmetric structure and distribution of the gains and losses across developed and developing countries associated with each of these two policy dimensions. Thus, global cooperation will be much easier to sustain when pursued through linked negotiations. This needs to be advocated more by new empirical and theoretical research.

The energy intensity of trade has not received adequate attention in the literature. Given that many of the environmental problems have energy use as their main source, the trade-energy-environment linkage needs to be disclosed. This is very important because the energy intensity of trade has been estimated to be higher than the energy intensity of overall economic activity, and than that of the transportation. In addition to the use of energy in the production of traded goods, the use of energy associated with the transportation of these goods internationally (an essential part of any trade activity) is even higher. It may be desirable to put new requirements in trade agreements in order to reduce the energy burden associated with the transportation of traded goods and services. More research is needed in this area; by cooperating with researchers from transportation economics, better outcomes may be obtained.

The impact of trade liberalization on the stringency of environmental regulations is another area, which is not studied adequately. The two competing views that, with trade liberalization, higher environmental standards of developed countries will be imported to developing countries and that competitiveness pressures will force countries with higher environmental standards to switch to more lenient environmental regulations ("race-to-the-bottom") need to be studied in more detail. A number of theoretical papers present mixed evidence, and there are only a few empirical papers on this subject. New empirical studies will be very useful; however, this requires a profound measurement of environmental policy stringency across countries, which is by itself a very demanding task.

Environmental quality is a normal good and demand for higher environmental quality will rise with increases in income. The current environmental awareness in developed countries is mainly due to increased income levels. The role of trade liberalization in reaching this wealthy position can not be ignored. For developing countries, the policies generating higher economic growth (including trade liberalization) are therefore important from an environmental perspective; however, developing countries do not need to repeat the environmental mistakes of developed countries. It is wrong to assume that trade liberalization can only be achieved in the framework drawn by the current GATT rules. It will be much better if free trade targeting of the GATT is changed to sustainable free trade. Short term expansion of GATT articles for the protection of the environment and the transformation of GATT into GATE (General Agreement on Trade and the Environment) over the long term emerges as an important policy. The key factors in the design of GATE will be to generate a consensus for treating trade and the environment equally (i.e. do not favor one over the other without any scientific reason), and to allow for a functional trade system with utmost environmental quality. This should be an important item of the agenda of the next round of the GATT.

Studies on trade and environment are very valuable as both trade and environment are essential components of modern life. As stated very nicely in Ekins et al. (1994), "If the world's trading system were to collapse, doubtless much hardship and suffering would result. But, if the global environment were to collapse, the result would be much worse." Thus, we need policies that will not sacrifice environment for trade and trade for environment. It is very vital to develop the institutions and the framework for environmentally sustainable trade, and research towards this goal should be given priority and support.

II

IMPACT OF ENVIRONMENTAL REGULATIONS ON INTERNATIONAL COMPETITIVENESS

Chapter 9

INTERNATIONAL COMPETITIVENESS AND ENVIRONMENTAL REGULATIONS

In the second part of our exposition on trade and the environment, we will try to identify the impact of environmental regulations on the competitiveness of the firms regulated. This is a concern both for the developed and developing countries. Developed countries fear that higher environmental regulations at home may cause a relocation of domestic industries to countries with lax environmental regulations. On the other hand, concerns of developing countries stem from the harmonization of environmental standards across countries (albeit differences in incomes) and/or trade bans on their exports because of environmental standards. This subject is a hot topic nowadays, and we will try to explore different aspects in the following chapters. Both the empirical and theoretical literature will be summarized.

1. Introduction

We will start with the measurement of competitiveness. Jaffe et al. (1995) present a very concise introduction on the alternative indicators of "competitiveness" and here we will make use of it. As this study is a survey of the empirical literature investigating the impact of environmental regulations on the competitiveness, it fits our framework perfectly. Jaffe et al. (1995) describe three different indicators of competitiveness. The first one is associated with the change in net exports of the goods subject to stricter regulations, and with comparisons between net exports of these goods and others produced under less regulated conditions. Empirically this measure can be captured by the magnitude and significance of the variable indicating the effect of regulatory stringency in a regression explaining changes in net exports across industries. A second potential indicator has to do with the degree of shifts in the locus of

S. Alpay (ed.), Trade and The Environment, 111–120.

production of pollution-intensive goods from countries with stringent regulations towards those with less. The third indicator of competitiveness is related to the attractiveness of countries as a locus of foreign direct investment depending on the stringency of their environmental regulations. We will use these three different indicators of competitiveness in our summary of the empirical literature in the proceeding chapters.

Conventionally, it was argued that environmental regulations would lower the competitiveness of the firms being regulated as compared to those subject to lax environmental conditions. This argument was assumed to be robust to the type of the environmental regulations. Recently, this view has been challenged by a revisionist school. They argue that properly crafted environmental regulations (i.e., *incentive-based*) not only bring social benefits with it (like increased environmental quality, decline in health risks associated with pollution, etc.), but also can increase competitiveness of the firms being regulated as higher environmental standards can trigger innovation that may offset the compliance costs. This debate is more explicitly seen in a series of papers published in Fall 1995 issue of Journal of Economic Perspectives, one by Palmer, Oates and Portney, and the other by Porter and van der Linde. We can associate *conventional school* with Palmer, Oates and Portney (1995) paper, and *revisionist school* with Porter and van der Linde (1995). The views of these two schools will be explored in the next two sections.

2. Conventional School

By the very notion of profit maximization, the firm must have realized all the profitable opportunities and therefore environmental regulation, which is a constraint by itself, should not be expected to increase the profits of the firms being regulated[1]. If it were so, the firms would not have been realizing their profitable opportunities and thus they were operating inside their production possibilities. In brief, there are no $10 bills lying on the ground that need to be picked up. More specifically, it has been stated that environmental regulation imposes new significant costs on domestic firms, and thus, given the lax regulations abroad, they would experience a fall in their international competitiveness.

In their 1994 paper, Oates, Palmer and Portney wrote:

> There has been widespread concern that the increasing stringency of domestic environmental regulation will put home-based industries at a competitive disadvantage in the

[1] Palmer et al. (1995) state that "The model essentially formalizes the basic point that the addition (or thightening) of constraints on a firm's set of choices can not be expected to result in an increased level of profits."

> international marketplace (see, for example, the Business Roundtable (1993)). According to this view, the increased cost that accompany more stringent controls will mean that domestic firms must confront their competitors abroad subject to an "unfair" burden. Such concern has even led to some proposed legislation in the form of a bill that would have introduced "counterveiling levies" against foreign nations whose "exports" benefit from the cost advantages associated with lax environmental programs.

Many theoretical studies in this literature present evidence for this view. One can see, Pethig (1976), Siebert (1977), Yohe (1979) and McGuire (1982). These studies, however, did not take possible innovations that may come with stricter regulation into account as suggested by the revisionist school. Recently, Simpson and Bradford (1996) presented an extensive analysis of this debate by incorporating main views of revisionist school into their model. Their results can be interpreted as a further support for the conventional school (this study will be explored in more detail below).

Palmer, Oates and Portney (1995) accept the possibility that stricter regulations might result in innovation, and innovation might offsett the costs associated with new regulations, however, they say, this is very rare in practise. Moreover, even if regulations foster innovation, they state that regulations will harm competitiveness by crowding out funds for other potential (and possibly more productive) projects.

3. Revisionist School

This view, also known as the Porter hypothesis, is proposed by Porter (1991). Porter states that the trade-off between environmental stringency and international competitiveness comes from the "static" approach to the problem. In fact, if the analysis is carried through a "dynamic" framework[2], which mainly encompasses possibilities of innovation in technology, product, and processes, then there is room for improvement in international competitiveness as a result of stringent environmental regulation. It is true that environmental regulation increases the number of constraints that the firms are facing; however, it may motivate the firm towards innovation, which may offset the costs associated with stricter regulations. Porter and van der Linde (1995) state:

> ... we will argue that properly designed environmental standards can trigger innovation that may partially or more than fully offset the costs of complying with them. Such "innovation offsets"... can not only lower the net cost of meeting environmental regulations, but can even lead to absolute advantages over firms in foreign countries not subject to similar regulations. ... In short, firms can actually benefit from properly

[2]The terms static and dynamic should not be considered in terms of a model that has an explicit time dimension, but rather in terms of the possibilities of innovation and other possible changes.

> crafted environmental regulations that are more stringent (or are imposed earlier) than those faced by their competitors in other countries. By stimulating innovation, strict environmental regulations can actually enhance competitiveness.

More explicitly, Porter hypothesis asserts that in general, the efficiency of an industry can be improved by more stringent environmental (or other) regulations either through sweeping inefficiencies out of the production process or through fostering innovation. As firms struggle to meet new environmental regulations, they reconsider their production processes and they can discover new techniques in abatement (and possibly production) technology; thus, it may be the case that overall, firms may experience a fall in their costs of production. Initially, these innovations have not been made because, facing lax environmental conditions, firms did not have incentives to invest in innovation activities.

A critical requirement of the Porter hypothesis is that environmental regulations are carefully crafted. Porter and van der Linde (1995) state that environmental regulations must have three characteristics in order to foster the innovation offsets that arise from new technologies and approaches to production. Firstly, regulations must facilitate opportunities for innovation; the regulatory agency should not dictate the approach to innovation, it must be left to the industry. Secondly, imposition of a particular technology must be avoided, and instead regulations should foster continuous improvement. Thirdly, regulations must be designed in a way that leaves as little room as possible for uncertainty at every stage. Porter and van der Linde suggest that "Environmental laws and regulations need to take three substantial steps: phrasing environmental rules as goals that can be met in flexible ways; encouraging innovation to reach and exceed those goals; and administering the system in a coordinated way".

Given that environmental regulations are properly crafted, then environmental regulations can serve at least six purposes:

> First, regulation signals companies about likely resource inefficiencies and potential technological improvements. Companies are still inexperienced in measuring their discharges, understanding the full costs of incomplete utilization of resources and toxicity, and conceiving new approaches to minimize discharges or eliminate hazardous substances. Regulation rivets attention on this area of potential innovation.
>
> Second, regulation focused on information gathering can achieve major benefits by raising corporate awareness. For example, Toxics Release Inventories, which are published annually as part of the 1986 Superfund reauthorization, require more than 20,000 manufacturing plants to report their releases of some 320 toxic chemicals. Such information gathering often leads to environmental improvement without mandating pollution reductions, sometimes even at lower costs.

Third, regulation reduces the uncertainty that investments to address the environment will be valuable. Greater certainty encourages investment in any area.

Fourth, regulation creates pressure that motivates innovation and progress. Our broader research on competitiveness highlights the important role of outside pressure in the innovation process, to overcome organizational inertia, foster creative thinking and mitigate agency problems. Economists are used to the argument that pressure for innovation can come from strong competitors, demanding customers or rising prices of raw materials; we are arguing that properly crafted regulation can also provide such pressure.

Fifth, regulation levels the transitional playing field. During the transition period to innovation-based solutions, regulation ensures that one company cannot opportunistically gain position by avoiding environmental investments. Regulations provide a buffer until new technologies become proven and learning effects reduce their costs.

Sixth, regulation is needed in the case of incomplete offsets. We readily admit that innovation cannot always completely offset the cost of compliance, especially in the short term before learning can reduce the cost of innovation-based solutions. In such cases, regulation will be necessary to improve environmental quality.

Stringent regulation can actually produce greater innovation and innovation offsets than lax regulation. Relatively lax regulation can be dealt with incrementally and without innovation, and often with "end-of-pipe" or secondary treatment solutions. More stringent regulation, however, focuses greater company attention on discharges and emissions, and compliance requires more fundamental solutions, like reconfiguring products and processes. While the cost of compliance may rise with stringency, then, the potential for innovation offsets may rise even faster. Thus the net cost of compliance can fall with stringency and may even turn into a net benefit.

Porter and van der Linde (1995).

An important criticism of conventional school against revisionist school is why firms do not take all profitable opportunities before the regulations; i.e. if they can increase their profits by innovation or other means, why do they wait for environmental regulations? In the next chapter, we will provide an answer to this question by demonstrating that the regulatory policy causes changes in some parameters, like permit prices and price elasticity of demand for the good, which are outside the control of the firm but have impact on its decision. We show that some non-feasible R&D projects can become profitable after the enactment of stricter regulations because of the associated changes in these parameters; but, firstly, a brief exposition to the linkages between (i) environmental-regulations and innovation, and (ii) innovation and competitiveness will be very useful. We do this in the next section.

4. Regulations, Innovation and Competitiveness

We will analyze principal components of the Porter hypothesis empirically. The first component is associated with whether stringent environmental regulations lead to innovation. We will refer to Jaffe and Palmer (1997) for the empirical assessment of this component. The second component is the impact of innovation on the competitiveness. Not much discussion is relevant here, and a strong consensus regarding the positive impact of innovation on the competitiveness exists. Nevertheless, we will present some empirical assessment of this view as well; our focus will be on policies to promote competitiveness through innovation.

5. Environmental Regulations and Innovation

A requirement of the Porter hypothesis is that environmental regulations are properly crafted, that is to say, they are incentive-based. Thus, to be able to assess the Porter hypothesis empirically, we first need to test whether incentive-based environmental regulations (and not any type of environmental regulation) lead to innovation. Not many empirical studies exist in the literature linking environmental regulations and innovation. Lanjouw and Mody (1996) investigate the impact of environmental compliance costs on the patenting of environmental technologies. They use cross-country data on environmental patents and expenditures for compliance with environmental regulations. Their estimations provide evidence for a positive correlation between expenditures for compliance with environmental regulations and environmental patenting, a support for the connection between regulations and innovation indicated in the Porter hypothesis. Jaffe and Palmer (1997) extend this study in two respects. They incorporate the impact of regulations on R&D expenditures (as well as on patenting), and they consider aggragate innovative activity instead of innovation only in environmental technology. They use US data.

Jaffe and Palmer (1997) divide their analysis into two parts. Firstly, they analyze the relationship between stringency of environmental regulations and R&D investment (i.e. innovative activity) by manufacturing firms through the following regression equation:

$$Log(R\&D)_{it} = \beta_1 log(VA)_{it} + \beta_2 log(R\&D_G)_{it} + \beta_3 log(PACE)_{it-1} + \alpha_i^R + \mu_t^R + \varepsilon_{it}^R$$

where i denotes industries, t denotes years, R&D is industry funded R&D expenditures, VA is industry value added, $R\&D_G$ is a proxy for government-funded R&D within the industry, and $PACE$ is pollution control expenditures from the Census Bureau's Pollution Abatement Costs and Expenditure survey. Current R&D is assumed to be affected by lagged regulatory stringency, and thus, lagged value of $PACE$ is used. Value added is included because R&D and $PACE$ vary with respect to the size of the industry. Time and industry dummies (μ_t^R and α_i^R, respectively) aim to remove the effects of industry variation – industries' technological characteristics – and time effects like, inflation and tax changes.

The environmental compliance cost data comes from the PACE survey, which has been conducted annually since 1973 in the US. The survey covers manufacturing firms and collects information on both the capital and operating costs of complying with environmental regulations. Value added data comes from US Census of Manufacturing. The R&D data comes from a survey done by the Census for the National Science Foundation (NSF). Government R&D is measured as the employee-weighted fraction of firms in the industry that report receiving government research funding.

Jaffe and Palmer show that government R&D variable has a significant and positive coefficient. The critical coefficient in this model is that of the $PACE$ variable; the coefficient of $PACE$ is positive and significant for the fixed-effect estimation. Thus, Jaffe and Palmer (1997) identify a positive correlation between the pollution-abatement expenditure (a proxy for the level of environmental regulations) and R&D investment. This result presents a partial support for the Porter hypothesis, in the sense that higher regulations will lead to higher abatement expenditure, and so higher R&D investment; however, what is more important whether investment in R&D leads to innovation (in the sense of new output or reduction in production costs). They make this connection in the second part of their analysis by assuming that patents are proportional to innovative output. The impact of environmental regulations on patents is identified through the following regression equation:

$$Log(PA)_{it} = \gamma_1 log(VA)_{it} + \gamma_2 log(FP)_{it} + \gamma_3 log(PACE)_{it-1} + \alpha_i^P + \mu_i^P + \varepsilon_{it}^P$$

where PA represent successful US patent applications in year t by US corporations, and FP (foreign patents) represents successful US applications in year t by foreign corporations. The data for the patent analysis is an industry panel of US patents by year of application. Note that the data on patents by industry are only a crude measure of inventive output by that industry.

Jaffe and Palmer (1997) find that both foreign patenting and domestic value added have positive, and generally significant coefficients. The critical finding is that the impact of lagged $PACE$ variable on the patents is not significant.

Thus, they state that regulatory compliance costs have no detectable impact on patenting activity. An overall empirical support regulation-innovation component of the Porter hypothesis could not be obtained. In brief, they conclude:

"Increased stringency of environmental regulation spurs increased innovative activity by firms. We found significant and positive relationship between regulatory compliance expenditures and R&D expenditures while there is no significant relationship between regulatory compliance expenditures and patenting activity... Whether or not regulation-inspired R&D leads to lower costs of production or new and improved products in the future remains an unanswered question."

We should remind that Jaffe and Palmer (1997) data is associated with US regulations, which are mostly command-and-control type, i.e. not incentive based; so, their results can not constitute a genuine test of the Porter hypothesis in the mentioned context[3]. Nevertheless they present an interesting framework which can be made use of with data on incentive based regulations. Finally, for a very recent extensive survey of interactions between environmental policy and technological change, one can see Jaffe et al. (2002).

6. Innovation and Competitiveness

In this section, we present a brief introduction to interactions between innovation and competitiveness. We will start with some empirical evidences and finish with some policy guidelines.

Mytelka (1999) presents interactions among competition, innovation and competitiveness through some case studies on four industrial sectors (namely machine tool industry, telecommunications, pharmaceutical and petrochemicals) in four countries (Brazil, Korea, Chinese Taipei, and India). During the 1970s and 1980s entry into machine tools and pharmaceuticals industries was relatively easy in Chinese Taipei due to low investment costs in the former and domestic protection in the latter. Although the two sectors faced similar conditions, machine tool industry and pharmaceuticals showed significant differences in their innovativeness. Machine tool industry, with higher innovative capability, had much better export performance.

---

[3]This is also valid for Lanjouw and Mody (1996) study mentioned above as past environmental regulations in many countries have been in the form of command-and-control.

As regards to cross-country comparisons, Mytelka states that Chinese Taipei machine tool firms were highly competitive as compared to their Indian counterparts because of the differences in the innovative strategies. Japanese model of innovation implemented in Chinese Taipei firms was the key element in their lead over India. In case of bio-pharmaceuticals, the picture is just the opposite. Through in-house R&D, the Indian pharmaceutical firms had built up their export capacity over the 1980s, and they lead over Chinese Taipei firms at investing in bio-pharmaceutical research in 1990s.

In another case study, Mytelka mentions Brazil's and Korea's strategies to develop their telecommunications equipment industries in the 1980s. This was done initially behind protective barriers, and governments reduced the risks of investment at the beginning in both countries. Carefully designed reforms in regards to domestic restructuring and deregulation prior to openning up the market to international competition, gave the opportunity to Korean firms to change their traditional ways of operation. This kind of transition together with public sector R&D support stimulated innovative capabilities in Korean firms. In Brazil, radical deregulation was combined with rapid market openning; adjustment of Brazilian firms to strong international competition was not possible. Korean producers achieved to cut their costs to comply with international prices and were able to export, due to their acquired innovative capability during the transition period while Brazilian firms failed in showing satisfactory innovation performance, and so became uncompetitive.

Mytelka (1999) concludes that the links between competition and competitiveness are more complex than suggested in the earlier works in this literature: "The argument, simply put, is that while competition may be necessary for the achievement of competitiveness, it is not sufficient. Nor is it sufficient, from a dynamic perspective, in maintaining competitiveness. What is also usually needed are innovative processes which will lead to the improvement of processes, products, management and organizational routines. Equally important, these innovative processes themselves have complex determinants that also need to be understood if the ultimate objective of increased competitiveness is to be achieved". Among the alternative stimulators of innovation, we can mention industry characteristics, requirements of suppliers or clients, and government policies. All these may in fact be linked to the Porter hypothesis. New environmental regulations may stimulate the kind competitive pressures that lead to innovation and competitiveness. Evidences for such a linkage have been shown for Asian countries for a number of government policies such as imposition of export performance criteria (see for example, Rodrik 1994). Similarly, requirement of better environmental performance by clients may induce firm towards innovation and, thus, make them more competitive.

It is suggested that an innovation-based approach to policy making would require policy makers to take two factors into consideration in the design of policies, which are aimed at the promotion of competitiveness. These are (i) competitive conditions across and within sectors over time, and (ii) the practices of economic actors with respect to competition and innovation. The sequencing and timing of the policies have also great importance. In regards to openning up the market to international competition, it is proposed that before the introduction of trade liberalization policies, domestic firms should have passed through a period of technological-capability-increasing followed by policy changes which stimulate the development of habits and practices of innovation, if market opening is to stimulate innovation and, thus, competitiveness.

A general conclusion in this strand of the competitiveness literature is that there are no simple rules for policy makers to follow to achieve the objective of international competitiveness; some rough guidelines can be thought of. "Drivers" of competitiveness can differ across industries and time; industry characteristics have to be identified carefully by the cooperation of policy-makers, bureaucrats, and representatives from private sector. Among the alternative types of innovation, the right one must be determined and stimulated for improved competitiveness. If the cost of production is the key element of competitiveness then process innovations need to be stimulated, and if product quality and differentiation are more important then product innovation should be the target. Finally, there is nothing wrong in thinking that environmental regulations may serve as stimulus for innovation if the policy makers design their policies in accordance with these guidelines. Although the objective of environmental policies is to improve environmental conditions and not the competitiveness, the outcome can be a win-win if the policies incorporate stimulus for innovation as suggested by the Porter hypothesis.

Chapter 10

COMPETITIVENESS AND ENVIRONMENTAL REGULATIONS: A THEORETICAL INQUIRY

1. Introduction

In this chapter, we will briefly introduce the theoretical literature on the impact of environmental regulations on the competitiveness of firms. Then, we will explore the relationship among environmental regulations, innovation and (international) competitiveness theoretically when the environmental agency uses *tradable emission permits* regulation system. In such a regulation system, incentives for innovation is larger since firms can earn revenues by selling extra permits. Thus, it is of interest to determine the impact of the parameters of permit market, such as price elasticity of permit demand, on the regulation–competitiveness debate. In so doing, we will also identify how the features of innovation process, like cost of and returns to innovation, and the probability of successful innovation, affect post-regulation competitiveness of the firms subject to higher regulation.

2. Theoretical Literature

Theoretical work, relatively limited compared to empirical work, is in favor of conventional school; see for example, Pethig (1976), Siebert (1977), Yohe (1979), McGuire (1982), Conrad (1993), Barrett (1994b), Rauscher (1994), and Ulph (1996). Rauscher (1994) presents an in depth analysis of "ecological dumping " phenomenon. Among the alternative definitions, one can characterise ecological dumping as a situation in which governments relax environmental standards in order to generate competitive advantage for domestic firms in international markets. He then shows that strategic trade policy and lobbying activities of the exporters are the main causes of ecological dumping. Barrett

S. Alpay (ed.), Trade and The Environment, 121–141.

(1994) shows that a "weak" environmental standard by domestic government may increase the competitiveness of a monopoly when the foreign industry is imperfectly competitive; if the domestic industry consists of more than one firm, this time there are incentives for "strong" standards[1]. Ulph (1996) demonstrates that governments have an incentive to distort their environmental regulations downwards from the optimal levels in order to lower the costs of the home firm, and thus increase their profits. Although these studies look at the issue from a reverse angle (they consider the impact of trade on environmental regulations), their implicit implication is that stricter environmental regulations will affect competitiveness of the firms negatively.

However, previous studies mentioned above ignore innovation incentives associated with stricter regulations, which is an important part of the Porter hypothesis. Simpson and Bradford (1996) include innovation incentives in their modeling. They consider a model of strategic trade a la Brander and Spencer (1985). Their model includes two countries with one firm in each. Governments regulate environmental conditions through an effluent tax. There are two different decisions to be made by firms: how-much to invest in innovation expenditure and how-much to produce. Innovation expenditure decision depends on the level of effluent tax. The domestic and foreign firms are assumed to compete in a third country's market to avoid consumer surplus considerations in the analysis of the model. Applying the usual backward solution procedure, the model is analyzed in three stages. In the third stage, firms determine their output levels as a function of effluent tax and innovation expenditure. In the middle stage, firms choose their cost-reducing innovation expenditures. In the first stage, governments identify effluent tax level by maximizing firm's profits net of its innovation investment and tax charges, less environmental damage plus tax receipts.

Simpson and Bradford show that it is possible to develop a model in which welfare maximization results in setting effluent taxes above the marginal damage of pollution. However, these situations are rare exceptions. They state that the impact of stricter environmental regulation on the performance of the industries being regulated would likely differ across sectors and be impossible to predict with any precision for any, but, that it is difficult to construct examples in which tougher regulation should be enacted to enhance the long-run competitiveness of domestic industry and thus "tightening regulation to induce advantage may be extremely dubious as practical policy advice."

[1] Strong (weak) standards mean that the marginal damage from pollution is less (higher) than marginal cost of abatement.

Some very recent studies on the impact of regulations on the competitiveness provide evidences for the Porter hypothesis. Xepapadeas and de Zeeuw (1998) show that downsizing and modernization of firms subject to environmental policy will increase average productivity, and will have positive effects on the marginal decrease of profits and environmental damage. They conclude, rather neutrally, that a win-win situation as suggested in the Porter hypothesis can generally not be expected, but the trade-off between environment and competitiveness is not so grim as is often suggested.

One final interesting side of the regulation–competitiveness debate is the impact of the *type* of the regulation and the *structure* of the market being regulated. Sartzetakis and Constantatos (1995) investigate how a country's choice of environmental policy tool affects international competitiveness of its firms. They show that total market share of regulated firms will be higher in case of tradable emission permits than in case of command and control due to better allocation of total abatement among the firms in the former. On the other hand, Ulph (1990) demonstrates that if trade is modelled as a one-shot Cournot equilibrium, countries do not have preference over the type of environmental policy, and if trade is modelled as a Stackelberg equilibrium, then both countries are better-off (in terms of producer surplus) if the follower uses standards (i.e. command and control).

3. Porter Hypothesis and Tradable Emissions Permits

In this part, we examine the impact of environmental regulations on the competitiveness of the regulated firms in a two-country, one-good model. We, in particular, try to see whether one can obtain theoretical evidences for the Porter hypothesis without imposing strong assumptions.

Similar to the work of Sartzetakis and Constantatos (1995), we assume that two Cournot oligopolists produce a good that generates pollution. To concentrate on the effect of regulation on the competitiveness, we will try to make simplifying assumptions where necessary. So, marginal cost of production, c, is assumed to be the same across firms and constant for analytical simplicity. Each firm maximizes its own profit taking the other firm's production decision into account. Inverse demand function is given by $p = a - bQ$, $a > 0, b > 0$, where p and Q represent price and demand of the good; supply of the good is $q_1 + q_2$, sum of the production of two firms, and so, $Q = q_1 + q_2$ in equilibrium.

Since the production process generates pollution, the firms are required to take some abatement actions. The emissions are proportional to the firms' output levels, and so $E_i = rq_i$, where E_i is the emission demand of firm i and

$r > 0$. Firms can reduce their emission levels either by decreasing the output level or by undertaking some abatement. Total abatement is represented by $A_i = \alpha_i q_i$, where α represents the abatement choice variable, with the corresponding convex abatement cost, $C_i = e_i A_i^2$, where e_i is a positive number.

The environmental agency implements tradable emission permits (TEP hereafter) regulation system to control aggregate emission level. As we stated above, the claim of revisionist school is that environmental regulation may increase competitiveness of the firm if environmental regulation is of the right type, right type being the incentive-based regulation as opposed to standard setting. TEP is one example of incentive-based instruments, so it can be used to analyze Porter's hypothesis. In this system, firms must have a prespecified amount of permits to discharge a certain amount of pollution; for example, they may need to surrender one permit per unit of emissions. Firms can trade permits among themselves. A thorough analysis of tradable emission permits can be found in Montgomery (1972).

Overall pollution is a result of many polluting industries inside the country. We assume that environmental regulation is applied to all polluting industries. We further assume that market for tradable permits is perfectly competitive and thus firms take the permit prices, P^ϵ, that causes zero net demand for the permits as given. Initially, a total of $\bar{E}_T$ amount of permits are distributed equally among firms (each is given $\bar{E}$). Firm i's demand for permits is given by the difference between its desired level of emissions and abatement less of initial permit endowment, i.e. $E_i - A_i - \bar{E}$.

Thus, the optimization problem of each firm is as follows:

$$\max_{q_i, \alpha_i} \quad \pi_i = pq_i - cq_i - e_i A_i^2 - P^\epsilon \left(E_i - A_i - \bar{E}\right) \tag{10.1}$$

Firms choose output and abatement level (through α) to maximize their profits. Replacing A_i and E_i in (10.1), we get an open form for firms' optimization problem:

$$\max_{q_i, \alpha_i} \quad pq_i - cq_i - e_i(\alpha_i q_i)^2 - P^\epsilon \left(rq_i - \alpha_i q_i - \bar{E}\right) \tag{10.2}$$

where α_i is the choice variable related to optimal abatement level. As shown in appendix, the reaction function for firm i is given by:

$$q_i = \frac{a - c - rP^\epsilon}{2b} - \frac{1}{2} q_j \tag{10.3}$$

Solving for optimal output levels, we get,

$$q_i = \frac{a - c - rP^\epsilon}{3b} \tag{10.4}$$

Firms produce the same level of output, which is not surprising under our symmetry assumptions[2]. Similarly the optimal abatement level is:

$$A_i = \alpha_i q_i = \frac{P^\epsilon}{2e_i} \tag{10.5}$$

The good will sell for,

$$p = a - b(q_1 + q_2) = \frac{a + 2c + 2rP^\epsilon}{3} \tag{10.6}$$

Upon the substitution of optimal values, corresponding profit levels will be:

$$\pi_i = \frac{(a - c - rP^\epsilon)^2}{9b} + \frac{(P^\epsilon)^2}{4e_i} + P^\epsilon \bar{E} \tag{10.7}$$

These values represent the positions of the firms before the stringent regulation program has been taken. It can be seen easily that the firm with lower abatement costs (i.e lower e) will have higher profits than the other firm. A change in P^ϵ, permit price, has ambiguous effects on the level of profits.

Now, let us see the impact of stricter environmental policies on the competitiveness of the firms.

3.1 Stricter Regulation, Innovation and Competitiveness

Environmental agency limits the number of permits from $\bar{E}_T$ to $\tilde{\bar{E}}_T$. As environmental regulation becomes more stringent, there are two options for the firms: status quo or innovation; that is to say, they may continue with their current production and abatement technology or they try to obtain a better abatement (and/or production) technology through increased investment in R&D. So, as a result of stringent regulation, with two firms, there may be three different responses:

Both firms choose to invest in R&D, or both continue without investing in R&D, or one of them decides to invest while the other does not. Since innovation

[2] If we had assumed a different average emission rate, r, across firms, then the firm with smaller emission rate would have produced more.

is an important component of the regulation–competitiveness discussion, in this section, we will only consider the cases where at least one firm invests in R&D.

Note that in this section, we concentrate on domestic economy, and set up a framework through which the regulation and innovation process can be studied. Then, once we have this set-up, we can extend it easily to analyze the relationship between international competitiveness and environmental regulation.

We assume that if the firm engages in R&D then it has to incur a cost of C_R and in return, with probability γ, it can get a new (abatement) technology, with parameters $\tilde{e}$, and $\tilde{r}$ where $\tilde{e} < e$, and $\tilde{r} < r$. Porter's view that stricter regulation makes innovation more likely can be formalized, for example, by letting γ a function of number of available permits; so we let $\gamma = \gamma(E)\,,\ \ \gamma'(E) < 0$, where E represents the amount of permits available to the firm. With probability $(1 - \gamma(E))$, firm will not be able to innovate, but it will still incur the cost of R&D[3]. Also, note that C_R will be a function of $\tilde{e}$, $\tilde{r}$, and γ.

A rise in the stringency of environmental regulations will impact permit market through increases in permit prices due to the decline in the total supply of permits. A change in permit prices naturally generates incentives for firms to change their decisions on the amount of production, level of abatement, and investment in R&D. We would like to note that changes in the permit market are *outside* the control of the firm; thus, this permit price externality associated with environmental regulation will change the value of the available projects to the firm.

Case I:

In this case, we assume that firm 2 engages in R&D, and firm 1 continues to use its old technology. Firm 2 is assumed to obtain a new technology with parameters $\tilde{r}$ and $\tilde{e_2}$ at a cost of C_R with probability $\gamma(\tilde{E})$. Also assume that permit demand is not very price-elastic and so P^ϵ rises to $\tilde{P}^\epsilon$. This will affect market shares, abatement and profit levels as follows (note that firm 2 will now maximize its expected profits):

$$\max_{q_2,\alpha_2}\ \gamma(\tilde{E})\left[pq_2 - cq_2 - \tilde{e}_2(\alpha_2 q_2)^2 - \tilde{P}^\epsilon\left(\tilde{r}q_2 - \alpha_2 q_2 - \tilde{E}\right) - C_R\right] + (1-\gamma(\tilde{E}))\left[pq_2 - cq_2 - e_2(\alpha_2 q_2)^2 - \tilde{P}^\epsilon\left(rq_2 - \alpha_2 q_2 - \tilde{E}\right) - C_R\right] \quad (10.8)$$

Firm 1's problem is same as above. As shown in appendix, solving these problems, we get:

[3]We assume away the positive spillover effects of innovation across firms as it is not a major component of this debate.

$$q_1 = \frac{a - c - 2r\tilde{P}^{\epsilon} + R_E\tilde{P}^{\epsilon}}{3b} \tag{10.9}$$

$$q_2 = \frac{a - c - 2R_E\tilde{P}^{\epsilon} + r\tilde{P}^{\epsilon}}{3b} \tag{10.10}$$

$$A_1 = \frac{\tilde{P}^{\epsilon}}{2e_1} \tag{10.11}$$

$$A_2 = \frac{\tilde{P}^{\epsilon}}{2(\gamma(\tilde{E})\tilde{e}_2 + (1 - \gamma(\tilde{E}))e_2)} \tag{10.12}$$

$$\pi_1 = \frac{\left(a - c - 2r\tilde{P}^{\epsilon} + R_E\tilde{P}^{\epsilon}\right)^2}{9b} + \frac{\left(\tilde{P}^{\epsilon}\right)^2}{4e_1} + \tilde{P}^{\epsilon}\tilde{E} \tag{10.13}$$

$$\pi_2 = \frac{\left(a - c - 2R_E\tilde{P}^{\epsilon} + r\tilde{P}^{\epsilon}\right)^2}{9b} + \frac{\left(\tilde{P}^{\epsilon}\right)^2}{4(\gamma(\tilde{E})\tilde{e}_2 + (1 - \gamma(\tilde{E}))e_2)} + \tilde{P}^{\epsilon}\tilde{E} - C_R \tag{10.14}$$

where $R_E = \gamma(\tilde{E})\tilde{r} + (1 - \gamma(\tilde{E}))r$. Derivation of these results are given in the appendix, in the proof of proposition 1 which we present now.

PROPOSITION 1 *When environmental regulation becomes more stringent, the firm that innovates will increase its market share and will undertake a higher abatement as compared to the one which does not innovate; however, the impact on the competitiveness is ambiguous.*

Proof of the above proposition is presented in the appendix. To summarize, the first part, ($q_2 > q_1$), follows from equations (10.9) and (10.10) with the reminder that $R_E < r$, i.e. expected return to R&D is better than old technology. The second part follows from the ambiguity in the relative magnitudes of π_1 and π_2.

The firm that engages in R&D benefits from the decline in emission level at each output level (decline in r), and that generates a reduction in the cost of production and causes an increase in its market share as compared to the other firm. Note that the change in permit prices does not affect this result since both firms will be facing the same prices.

With a better "expected" abatement technology, the innovative firm produces a higher abatement (equations (10.11) and (10.12)); this is so because of a decline in the marginal cost of abatement which is proportional to $(\gamma(\tilde{E})\tilde{e}_2 + (1-\gamma(\tilde{E}))e_2)$ [4]. Higher abatement increases permit revenues for the innovating firm; so, through this channel, firm's expected profits tend to increase; however, larger abatement imposes new costs on the firm. Also, the cost of R&D needs to be taken into consideration. Under certain conditions as shown in appendix in the proof of above proposition, environmental regulation increases the competitiveness of the firm which undertakes innovation; the status quo firm loses from stringent regulatory conditions. Nevertheless, a lower profit level for the innovating firm as compared to non-innovating one is not impossible, due to uncertainty associated with innovation process. This would be the case if the cost of the innovation is very high, and/or returns to innovation is not satisfactory (i.e. high $\tilde{r}$ and $\tilde{e}$), and/or probability of innovation is low. Also note that, a decline in the number of available permits generate mixed impacts: a fall in permit revenues (through the term $P^{\epsilon}\bar{E}$) and a rise in the probability of innovation, $\gamma(E)$ since $\gamma'(E) < 0$.

In this section, we focused on inter-firm comparison and identified certain features related to innovation. Although innovation increases market share of the innovating firm, there is not full evidence for increased competitiveness.

Case II

In this case, both firms are motivated by stringent regulation and both of them invest in R&D. As opposed to inter-firm comparison of the previous case, here, we compare the market shares, abatement and profit levels of the firms with their pre-regulation levels. Assuming that the probability of innovation is same for both firms, the optimal values for output, abatement, and profits will be as follows:

$$q_1 = q_2 = \frac{a - c - R_E\tilde{P}^{\epsilon}}{3b} \qquad (10.15)$$

$$A_i = \frac{\tilde{P}^{\epsilon}}{2(\gamma(\tilde{E})\tilde{e}_i + (1-\gamma(\tilde{E}))e_i)} \qquad (10.16)$$

$$\pi_i = \frac{\left(a - c - R_E\tilde{P}^{\epsilon}\right)^2}{9b} + \frac{\left(\tilde{P}^{\epsilon}\right)^2}{4(\gamma(\tilde{E})\tilde{e}_i + (1-\gamma(\tilde{E}))e_i)} + \tilde{P}^{\epsilon}\tilde{E} - C_R \qquad (10.17)$$

[4] We assume that this parameter is smaller than e_1 and it is certainly less than e_2.

With higher environmental standards, firms begin to abate more; this is not surprising since stringent regulation increases returns to abatement. Formally, it can be seen from the comparison of equations (10.16) and (10.5).

Comparison of post- and pre-regulation output and profit levels is not so clear. From equations (10.15) and (10.4), if $rP^{\epsilon} > R_E\tilde{P}^{\epsilon}$, then post-regulation output levels will be higher than pre-regulation ones[5]; so, depending on innovation parameters $(\gamma, \tilde{r}, \tilde{e})$ and price elasticity of permit market, stricter regulation might result in higher production.

Now, let us try to identify impact of stricter regulation on the competitiveness of the innovating firms.

PROPOSITION 2 *More stringent environmental regulation will result in higher (lower) profits for both of the innovating firms if* $rP^{\epsilon} > (<)R_E\tilde{P}^{\epsilon}$ *and if the cost of innovation satisfies*

$$\frac{\left(\tilde{P}^{\epsilon}\right)^2}{4(\gamma(\tilde{E})\tilde{e}_i + (1-\gamma(\tilde{E}))e_i)} + \tilde{P}^{\epsilon}\tilde{E} - \frac{(P^{\epsilon})^2}{4e_i} - P^{\epsilon}\bar{E} > (<)C_R.$$

Proof easily follows from the comparison of profit expressions in (10.17) and (10.7); note that first term of (10.17) is greater (smaller) than that of (10.7), since $R_E\tilde{P}^{\epsilon} < (>)rP^{\epsilon}$; similarly, the condition on the cost of innovation makes the comparison of other terms in (10.17) and in (10.7) possible.

A careful investigation, however, shows that not all firms can become more competitive with the enactment of stricter regulations; above proposition does not take the impact of stricter regulation on other industries into account. Thus, the cases where the profits of both firms increase with the enactment of new environmental regulations should not be interpreted as a full support for the Porter hypothesis[6]. To show this more explicitly, let us assume that there is only one industry in the economy, and permit trades can only occur among the two firms that are being examined here. Unless environmental agency buys extra permits from the regulated firms, at least one of the firms becomes less competitive with the enactment of higher environmental standards. If permit trade only occurs between the firms, either there is no trade or one of the firms buys permits and the other one sells them; thus, reducing higher abatement and regulation costs through permit revenues is possible *only* for one firm. Even

[5] A detailed explanation on the mentioned condition will be given below.

[6] Even without the following considerations, this proposition does not present an unconditional evidence for the Porter hypothesis as can be seen from its conditional nature.

though innovation might lower the marginal cost of abatement, total abatement cost will always be higher due to higher abatement associated with higher permit prices. Explicitly, post-regulation abatement cost is (from equation 10.16 and the definition of abatement costs)

$$(\gamma(\tilde{E})\tilde{e}_i + (1-\gamma(\tilde{E}))e_i)A_i^2 = \frac{(\tilde{P}^\epsilon)^2}{4(\gamma(\tilde{E})\tilde{e}_i + (1-\gamma(\tilde{E}))e_i)} .$$

This is higher than pre-regulation abatement costs, $\frac{(P^\epsilon)^2}{4e_i}$, which can easily be calculated from equation (10.5). Thus, the firm buying the permits, will experience a fall in its competitiveness due to higher abatement costs and permit purchases.

In this section, we analyzed the interactions between regulation, innovation and competitiveness in a domestic economy. Our results show that when environmental regulation becomes more stringent, the profits of the regulated firms may move in either direction depending on returns to, probability and cost of innovation, and also on the properties of the permit market. Now, we can extend these results to a more general case that includes international markets.

3.2 International Competitiveness and Regulation

Given this basic set-up for the interactions between regulation, innovation and competitiveness, now we can extend our results into open economy. Through this extension, we would be able to present evidences or counter evidences for the views of conventional and revisionist school, and also see how the interactions between regulation, innovation and competitiveness compare in a closed economy versus in an open economy. We start with the description of a simple open economy.

We assume two countries, home (H) and foreign (F). Each country produces the same good. The demand for the product is represented by[7]

$$p = a - b(Q^H + Q^F) .$$

As in the previous section, each country's output is produced by two Cournot oligopolists. Each country applies tradable emission permit regulation policy,

[7] For ease of exposition, here we assume that home and foreign production are perfect substitutes for each other; however, our results are independent of this assumption.

however the regulatory environment is relatively lax initially. We will investigate the effects of an increase in environmental controls in home country. Let us first determine the initial positions of the firms in each country. To make the analysis more tractable with four different firms, we assume that the firms are symmetric in every aspect inside and across the countries; this will help us to isolate the impact of environmental regulation on the competitiveness of the firms that are otherwise identical.

As shown in appendix, profit maximization results in following output, abatement and profit levels:

$$q_i^H = q_i^F = \frac{a - c - rP^\epsilon}{5b} \tag{10.18}$$

$$A_i^H = A_i^F = \frac{P^\epsilon}{2e} \tag{10.19}$$

$$\pi_i^H = \pi_i^F = \frac{(a - c - rP^\epsilon)^2}{25b} + \frac{(P^\epsilon)^2}{4e} + P^\epsilon \bar{E} \tag{10.20}$$

Initially, each country distributes $\bar{E}_T$ amount of permits among the firms. Since countries are assumed to be identical, output levels, abatement levels and profits are also identical. At such an initial equilibrium position, home country puts new environmental regulations into effect and decreases the number of tradable permits from $\bar{E}_T$ to $\tilde{\bar{E}}_T$. Then, we want to determine what will happen to the competitiveness of the domestic firms (i) if they both continue with their present technologies, (ii) if both of the firms try to innovate and get a better technology at some cost, and (iii) if one of the firms tries to innovate and the other one keeps its old technology. Let us investigate each case in turn:

Case I:

In this case, we assume that none of the domestic firms takes R&D. The market shares of the firms will remain the same, if we assume that price elasticity of permit demand is very elastic; otherwise, permit prices, P^ϵ will increase in home country and market shares of the domestic firms will fall and the share of foreign firms will rise; an expected result. Note that since environmental policy does not change in foreign country, permit prices will remain at P^ϵ. If the new permit price level at home is $\tilde{P}^\epsilon$, new output, abatement and profit levels will be as follows:

$$q_i^H = \frac{a - c + (2P^\epsilon - 3\tilde{P}^\epsilon)r}{5b} \tag{10.21}$$

$$q_i^F = \frac{a - c + (2\tilde{P}^\epsilon - 3P^\epsilon)r}{5b} \tag{10.22}$$

$$A_i^H = \frac{\tilde{P}^\epsilon}{2e} \tag{10.23}$$

$$A_i^F = \frac{P^\epsilon}{2e} \tag{10.24}$$

$$\pi_i^H = \frac{\left(a - c + (2P^\epsilon - 3\tilde{P}^\epsilon)r\right)^2}{25b} + \frac{\left(\tilde{P}^\epsilon\right)^2}{4e} + \tilde{P}^\epsilon \bar{E} \tag{10.25}$$

$$\pi_i^F = \frac{\left(a - c + (2\tilde{P}^\epsilon - 3P^\epsilon)r\right)^2}{25b} + \frac{(P^\epsilon)^2}{4e} + P^\epsilon \bar{E} \tag{10.26}$$

Market shares of foreign firms will exceed those of domestic firms; this follows from equations (10.21) and (10.22); since $\tilde{P}^\epsilon > P^\epsilon$, then $(2P^\epsilon - 3\tilde{P}^\epsilon) < (2\tilde{P}^\epsilon - 3P^\epsilon)$, and so $q_i^H < q_i^F$. Comparison of profit levels is not as straightforward. If permit demand is very price elastic so that $\tilde{P}^\epsilon \approx P^\epsilon$ then we will have $\pi_i^F > \pi_i^H$, which is a support for the view of conventional school. In general, $\tilde{P}^\epsilon > P^\epsilon$, and comparison of profits becomes more difficult. The inter-firm comparison can also be coupled with the comparison of the profits of firms across regulation stringency. This is very clear in foreign firms case and can be stated as follows:

PROPOSITION 3 *Stricter regulation at home country will make foreign firms better-off unconditionally.*

Since $P^\epsilon < \tilde{P}^\epsilon$, then $(2\tilde{P}^\epsilon - 3P^\epsilon) > (-P^\epsilon)$, and so the profits of foreign firms after stringent regulation (equation (10.26)) is greater than their profits before regulation (equation (10.20)). The difference in permit prices increases the cost of production for regulated domestic firms relative to foreign ones. This rises the share of foreign firms in total production[8]. Given the same abatement, permit trade and regulatory conditions abroad, the profits of foreign firms increase.

[8]One can also show that total production falls and so price of the good rises after regulation.

Now, let us continue with the more important case where regulated domestic firms take actions for innovation with the imposition of new regulations.

Case II:

In this case, domestic firms engage in R&D, and they succeed in innovation with a probability of $\gamma(\tilde{E})$. As above, new technology variables will be $\tilde{e}$ and $\tilde{r}$. Also, permit prices in home country rise to $\tilde{P}^{\epsilon}$ with stricter regulation. The resulting values of output, abatement and profit levels for both domestic and foreign firms are as follows:

$$q_i^H = \frac{a - c + 2rP^{\epsilon} - 3R_E\tilde{P}^{\epsilon}}{5b} \tag{10.27}$$

$$q_i^F = \frac{a - c + 2R_E\tilde{P}^{\epsilon} - 3rP^{\epsilon}}{5b} \tag{10.28}$$

$$A_i^H = \frac{\tilde{P}^{\epsilon}}{2(\gamma(\tilde{E})\tilde{e} + (1 - \gamma(\tilde{E}))e)} \tag{10.29}$$

$$A_i^F = \frac{P^{\epsilon}}{2e} \tag{10.30}$$

$$\pi_i^H = \frac{\left(a - c + 2rP^{\epsilon} - 3R_E\tilde{P}^{\epsilon}\right)^2}{25b} + \frac{\left(\tilde{P}^{\epsilon}\right)^2}{4(\gamma(\tilde{E})\tilde{e} + (1 - \gamma(\tilde{E}))e)} + \tilde{P}^{\epsilon}\tilde{E} - C_R \tag{10.31}$$

$$\pi_i^F = \frac{\left(a - c + 2R_E\tilde{P}^{\epsilon} - 3rP^{\epsilon}\right)^2}{25b} + \frac{(P^{\epsilon})^2}{4e} + P^{\epsilon}\bar{E} \tag{10.32}$$

where $R_E = \gamma(\tilde{E})\tilde{r} + (1 - \gamma(\tilde{E}))r$. Then,

PROPOSITION 4 *When environmental regulation becomes more stringent at home and the domestic firms innovate after environmental regulatory change, market share of the domestic (foreign) firms will increase if* $rP^{\epsilon} > (<) R_E\tilde{P}^{\epsilon}$*; the firms in home country will undertake a higher level of abatement as compared to foreign country. Finally, domestic (foreign) firms will be more competitive if the following condition also holds*

$$\frac{\left(\tilde{P}^{\epsilon}\right)^2}{4(\gamma(\tilde{E})\tilde{e} + (1 - \gamma(\tilde{E}))e)} + \tilde{P}^{\epsilon}\tilde{E} - \frac{(P^{\epsilon})^2}{4e} - P^{\epsilon}\bar{E} > (<) C_R\,.$$

Proof is presented in the appendix.

Depending on the values of new technology parameters ($\tilde{e}$ and $\tilde{r}$), the cost of R&D (C_R), the probability of innovation (γ), price elasticity of permit demand, and permit price differential across countries, our model presents partial evidence for the view of revisionist school (partially, due to our previous discussion on the overall impact of regulation) or that of conventional school. It is possible that domestic firms not only increase their profits over the profits of foreign firms but also over their pre-regulation profits (shown in appendix). Moreover, post-regulation profits of foreign firms will be less than their pre-regulation profits if[9] $rP^{\epsilon} > R_E\tilde{P}^{\epsilon}$, and this follows from the comparison of profit expressions in (10.20) and (10.32). Relative magnitudes of r and $R_E = \gamma(\tilde{E})\tilde{r} + (1 - \gamma(\tilde{E}))r$, as well as $\gamma(\tilde{E})$, show the partial impact of innovation on this debate.

How does this case compare to the one in which only one of the domestic firms innovates? Will the competitiveness of the only innovating firm be higher than this case? We take up these questions now.

Case III

This case is associated with R&D by firm 1 and status quo by firm 2 in home country. Firm 1 maximizes its expected profits. The resulting output, abatement, and profit levels are as follows:

$$q_1^H = \frac{a - c + 2rP^{\epsilon} - 4R_E\tilde{P}^{\epsilon} + r\tilde{P}^{\epsilon}}{5b} \tag{10.33}$$

$$q_2^H = \frac{a - c + 2rP^{\epsilon} + R_E\tilde{P}^{\epsilon} - 4r\tilde{P}^{\epsilon}}{5b} \tag{10.34}$$

$$q_1^F = q_2^F = \frac{a - c - 3rP^{\epsilon} + R_E\tilde{P}^{\epsilon} + r\tilde{P}^{\epsilon}}{5b} \tag{10.35}$$

$$A_1^H = \frac{\tilde{P}^{\epsilon}}{2(\gamma(\tilde{E})\tilde{e} + (1 - \gamma(\tilde{E}))e)} \tag{10.36}$$

[9] If we reverse this assumption, the profits of the foreign firms will increase with stricter regulation as in previous proposition.

$$A_2^H = \frac{\tilde{P}^\epsilon}{2e} \tag{10.37}$$

$$A_i^F = \frac{P^\epsilon}{2e} \tag{10.38}$$

$$\pi_1^H = \frac{\left(a - c + 2rP^\epsilon - 4R_E\tilde{P}^\epsilon + r\tilde{P}^\epsilon\right)^2}{25b} + \frac{\left(\tilde{P}^\epsilon\right)^2}{4e_e} + \tilde{P}^\epsilon\tilde{E} - C_R \tag{10.39}$$

$$\pi_2^H = \frac{\left(a - c + 2rP^\epsilon + R_E\tilde{P}^\epsilon - 4r\tilde{P}^\epsilon\right)^2}{25b} + \frac{\left(\tilde{P}^\epsilon\right)^2}{4e} + \tilde{P}^\epsilon\tilde{E} \tag{10.40}$$

$$\pi_1^F = \pi_2^F = \frac{\left(a - c - 3rP^\epsilon + R_E\tilde{P}^\epsilon + r\tilde{P}^\epsilon\right)^2}{25b} + \frac{(P^\epsilon)^2}{4e} + P^\epsilon\bar{E} \tag{10.41}$$

where $e_e = (\gamma(\tilde{E})\tilde{e} + (1 - \gamma(\tilde{E}))e)$, and R_E is defined above.

PROPOSITION 5 *The non-innovating domestic firm will be less competitive when the other firm innovates after regulation. If marginal cost of emissions falls for the innovating domestic firm then (i) that firm will produce more than foreign firms, and if the cost of innovation is low, its post-regulation profits will be higher than all other firms, and (ii) the returns to innovation (as measured by the level of profits) will be higher when only one firm innovates than when both firms innovate.*

The first part of the proposition is easy and can be seen by comparing post-regulation profits of firm 2 when there is no innovation (equation 10.25) with its profits here (equation 10.40) by noting that $\tilde{P}^\epsilon > P^\epsilon$ and $R_E < r$. Remaining parts are proven in the appendix. The last part is not surprising as when there is only one innovating firm, the demand for its permits will be higher than when both firms innovate (and abate more); however, higher permit prices also increase unit cost of production (due to higher emissions cost); the overall impact on profits is conditional.

3.3 Summary

We present a framework where we can identify the factors that relate to the impact of stricter environmental policies on the international competitiveness of the regulated firms. Our study incorporates tradable emission permits as environmental regulation system, and this turns out to be an important factor on this debate.

The very notion of trade in emission permits opens up new opportunities and thus abatement and innovation become more attractive for the regulated firms. It is shown that innovation always results in higher market share as compared to non-innovating firms subject to regulation; however, profit levels, which measure the competitiveness, may or may not respond positively to regulation due to uncertain innovation prospects. It is possible that innovating firms become more competitive after regulation, but, when every regulated firm is taken into consideration, overall impact of regulation should not be expected to conform to the claims of the Porter hypothesis. This has been shown formally under the assumptions that firms can not sell their extra permits back to environmental agency and production technology is not affected by innovation.

It has been explicitly demonstrated how the parameters of innovation process, such as cost of, probability of and returns to innovation, and the features of the permit market, such as price elasticity of permit demand and permit price differentials across countries, have impact on the competitiveness of the regulated firms when the regulated firms respond to regulations by either investing or not investing in R&D.

Previous research concentrated on the impact of environmental regulations on the supply-side of the economy. We suggest that changes in environmental policies may also affect demand-side of the economy. If consumers have preference for the good that is produced in a relatively cleaner way, then imposition of higher environmental standards at home can increase the demand for domestic production, and thus, it opens up a channel towards increased competitiveness. This has to be shown formally in an extended version of the present model.

3.4 Appendix

Let us begin with the problem of each firm. Firms maximize their profits by choosing level of production and abatement; substituting p by $a - b(q_i + q_j)$ in (10.2):

$$\max_{q_i, \alpha_i} \quad (a - bq_i - bq_j)q_i - cq_i - e_i(\alpha_i q_i)^2 - P^\epsilon \left(rq_i - \alpha_i q_i - \bar{E}\right) \quad (10.42)$$

Differentiate (10.42) with respect to q_i:

$$a - 2bq_i - bq_j - c - 2e_i(\alpha_i)^2 q_i - P^\epsilon(r - \alpha_i) = 0 \quad (10.43)$$

and with respect to α_i:

$$q_i(-2e_i\alpha_i q_i + P^\epsilon) = 0 \quad (10.44)$$

From equation (10.44),

$$A_i = \alpha_i q_i = \frac{P^\epsilon}{2e_i} \quad (10.45)$$

which is equation (10.5). Using equation (10.45) in equation (10.43), we obtain

$$q_i = \frac{a - c - rP^\epsilon}{2b} - \frac{1}{2}q_j$$

which is the reaction function in (10.3). Similarly,

$$q_j = \frac{a - c - rP^\epsilon}{2b} - \frac{1}{2}q_i$$

Replacing this in the previous equation, we get the optimal output level in (10.4).

Proof of Proposition 1:

After replacing p by $a - b(q_1 + q_2)$, differentiate (10.8) with respect to q_2:

$$a - 2bq_2 - bq_1 - c - 2[\gamma(\tilde{E})\tilde{e_2} + (1 - \gamma(\tilde{E}))e_2]\alpha_2^2 q_2 - \tilde{P}^\epsilon[\gamma(\tilde{E})\tilde{r} + (1 - \gamma(\tilde{E}))r - \alpha_2] = 0 \quad (10.46)$$

and with respect to α_2:

$$q_2[-2(\gamma(\tilde{E})\tilde{e}_2 + (1 - \gamma(\tilde{E}))e_2)\alpha_2 q_2 + \tilde{P}^\epsilon] = 0 \quad (10.47)$$

To get the output and abatement levels, we also need to solve for the optimization problem of firm 1; but, this has already been shown above, in the first part of the appendix, as firm 1 does not engage in R&D. Solving equations (10.43), (10.44), (10.46) and (10.47) together, we get optimal output and abatement levels in equations (10.9)–(10.12). Using these values one can get the corresponding profit levels.

Since $\tilde{r} < r$, $\gamma(\tilde{E})\tilde{r} + (1-\gamma(\tilde{E}))r < r$, (note that $\gamma(\tilde{E})$ is between 0 and 1 since it measures the probability of innovation) and so, $-2(\gamma(\tilde{E})\tilde{r} + (1 - \gamma(\tilde{E}))r) + r > -2r + (\gamma(\tilde{E})\tilde{r} + (1-\gamma(\tilde{E}))r)$, hence, $q_2 > q_1$. Also, since $\gamma(\tilde{E})\tilde{e_2} + (1-\gamma(\tilde{E}))e_2 < e_2$, then $A_2 > A_1$. However, profit levels can not be compared without further specifications. Using the profit expressions in (10.13) and (10.14), if we assume that

$$\frac{\left(\tilde{P^\epsilon}\right)^2}{4(\gamma(\tilde{E})\tilde{e}_2 + (1-\gamma(\tilde{E}))e_2)} - C_R > \frac{\left(\tilde{P^\epsilon}\right)^2}{4e_1} \longrightarrow$$

$$\left(\tilde{P^\epsilon}\right)^2 \frac{\left(e_1 - (\gamma(\tilde{E})\tilde{e}_2 + (1-\gamma(\tilde{E}))e_2)\right)}{4e_1(\gamma(\tilde{E})\tilde{e}_2 + (1-\gamma(\tilde{E}))e_2)} > C_R\,,$$

then $\pi_2 > \pi_1$. This condition is a sufficient condition, and not a necessary one. It clearly shows that the effect of regulation on the competitiveness of the firms not only depends on the parameters of innovation but also on the features of the permit market, as reflected in permit prices.

International Competitiveness:

First, let us try to obtain the optimal output and abatement levels when there are foreign firms competing in the same market with domestic firms. Firm i in the home country solves the following optimization problem:

$$\max_{q_i^H, \alpha_i^H} \; pq_i^H - cq_i^H - e_i(\alpha_i^H q_i^H)^2 - P^\epsilon\left(rq_i^H - \alpha_i^H q_i^H - \bar{E}\right) \tag{10.48}$$

where $p = a - b(Q_H + Q_F)$, and $Q_k = q_1^k + q_2^k$, $k = H, F$. First order conditions result in the following reaction functions:

$$q_i^H = \frac{a - c - rP^\epsilon}{2b} - \frac{1}{2}q_j^H - \frac{1}{2}Q_F \tag{10.49}$$

The same reaction functions apply in case of foreign firms, and so,

$$q_i^F = \frac{a - c - rP^\epsilon}{2b} - \frac{1}{2}q_j^F - \frac{1}{2}Q_H \tag{10.50}$$

Using equations (10.49), (10.50) and noting that $Q_H = q_1^H + q_2^H$ and $Q_F = q_1^F + q_2^F$, we get

$$q_i^H = q_i^F = \frac{(a - c - rP^\epsilon)}{5b} \tag{10.51}$$

which is same as equation (10.18). Abatement levels will be same as closed economy case. Using these results, one can easily obtain profit levels in equation (10.20).

Proof of Proposition 4:

Both of the domestic firms maximize their expected profits:

$$\max_{q_i^H, \alpha_i^H} \gamma(\tilde{E}) \left[pq_i^H - cq_i^H - \tilde{e}(\alpha_i^H q_i^H)^2 - \tilde{P}^\epsilon \left(\tilde{r}q_i^H - \alpha_i^H q_i^H - \tilde{E}\right) - C_R\right] +$$
$$(1 - \gamma(\tilde{E})) \left[pq_i^H - cq_i^H - e(\alpha_i^H q_i^H)^2 - \tilde{P}^\epsilon \left(rq_i^H - \alpha_i^H q_i^H - \tilde{E}\right) - C_R\right] \tag{10.52}$$

Optimization problem of foreign firms is same as above in equation (10.48). The reaction functions are now as follows:

$$q_i^H = \frac{a - c - R_E\tilde{P}^\epsilon}{2b} - \frac{1}{2}q_j^H - \frac{1}{2}Q_F \tag{10.53}$$

$$q_i^F = \frac{a - c - rP^\epsilon}{2b} - \frac{1}{2}q_j^F - \frac{1}{2}Q_H \tag{10.54}$$

Solving these equations, we get the optimal output and abatement and profit levels given in (10.27)–(10.32).

Market shares of the innovating domestic firms and foreign firms are given by equations (10.27) and (10.28). If $R_E\tilde{P}^\epsilon < (>) rP^\epsilon$, then $(2rP^\epsilon - 3R_E\tilde{P}^\epsilon) > (<)(2R_E\tilde{P}^\epsilon - 3rP^\epsilon)$, and so $q_i^H > (<) q_i^F$.

Higher abatement at home country follows from the fact that $\gamma(\tilde{E})\tilde{e} + (1 - \gamma(\tilde{E}))e < e$; see equations (10.29) and (10.30).

The competitiveness will be measured by the changes in the profit levels which we show now. We will consider the case where $R_E\tilde{P}^\epsilon < rP^\epsilon$, and

$$\frac{\left(\tilde{P}^\epsilon\right)^2}{4(\gamma(\tilde{E})\tilde{e} + (1 - \gamma(\tilde{E}))e)} + \tilde{P}^\epsilon\tilde{E} - \frac{(P^\epsilon)^2}{4e} - P^\epsilon\bar{E} > C_R\,;$$

by reversing the direction of inequalities, we get the alternative result in the proposition.

The first term in the profit expression of domestic firm in (10.31) is greater than the corresponding term in the profit expression of foreign firms in (10.32) since they are same as market shares (to the second power). If the cost of R&D satisfies the second condition in the proposition, then other terms also become larger in π_i^H than in π_i^F, and so $\pi_i^H > \pi_i^F$. Thus, domestic firms become more competitive after the enactment of stricter environmental regulation. Under given conditions, new profits of domestic firms (equation 10.31) will exceed their pre-regulation profits (equation 10.20) since $-rP^\epsilon < (2rP^\epsilon - 3R_E\tilde{P}^\epsilon)$.

Proof of Proposition 5:

In this case, firm 2, the innovating domestic firm, maximizes its expected profits (as in 10.52) and domestic firm 1 and foreign firms maximize their profits (as in 10.48). Solving the corresponding first order conditions, we first obtain the following reaction functions:

$$q_1^H = \frac{a - c - R_E\tilde{P}^\epsilon}{2b} - \frac{1}{2}q_2^H - \frac{1}{2}Q_F \tag{10.55}$$

$$q_2^H = \frac{a - c - r\tilde{P}^\epsilon}{2b} - \frac{1}{2}q_1^H - \frac{1}{2}Q_F \tag{10.56}$$

$$q_i^F = \frac{a - c - rP^\epsilon}{2b} - \frac{1}{2}q_j^F - \frac{1}{2}Q_H \tag{10.57}$$

then solving these together, we get equations (10.33)–(10.41).

Marginal cost of emissions measure the amount that needs to be paid for emissions for a unit increase in output. Before the stricter regulation, it is rP^ϵ and post-regulation it becomes $R_E\tilde{P}^\epsilon$. So, the condition in the proposition translates into $rP^\epsilon > R_E\tilde{P}^\epsilon$.

Market shares of the innovating domestic firm, firm 1, and the other firms are given by equations (10.33)–(10.35). Since $R_E < r$, then $(2rP^\epsilon - 4R_E\tilde{P}^\epsilon + r\tilde{P}^\epsilon) > (2rP^\epsilon + R_E\tilde{P}^\epsilon - 4r\tilde{P}^\epsilon)$, and so $q_1^H > q_2^H$. Similarly, $(2rP^\epsilon - 4R_E\tilde{P}^\epsilon + r\tilde{P}^\epsilon) > (-3rP^\epsilon + R_E\tilde{P}^\epsilon + r\tilde{P}^\epsilon)$, since $R_E\tilde{P}^\epsilon < rP^\epsilon$, and thus, $q_1^H > q_i^F$.

The first term in the profit expression of innovating firm in (10.39) is greater than the corresponding term in the profit expression of the non-innovating domestic firm and foreign firms, equations (10.40) and (10.41), respectively, since they are same as market shares (to the second power). If the cost of R&D satisfies the condition in the previous proposition (prop. 4), then the other terms also become larger in π_1^H than in π_i^F, and so $\pi_1^H > \pi_i^F$. Thus, as suggested by

Porter, the firm subject to stricter regulation can become more competitive by investing to innovate.

Finally, let us compare the returns to innovation when only one firm innovates versus when both firms innovate. This requires us to compare profits of innovating domestic firm in (10.39) with (10.31), profits of domestic firms when they both innovate. Since these expressions only differ in the first terms, we just need to compare the first terms. As $R_E\tilde{P}^\epsilon < rP^\epsilon$, then $(2rP^\epsilon - 4R_E\tilde{P}^\epsilon + r\tilde{P}^\epsilon) > (2rP^\epsilon - 3R_E\tilde{P}^\epsilon)$, and so π_1^H in (10.39), which represents the profits of the innovating firm when only one firm innovates, is greater than π_i^H in (10.31), which represents the profits of the two domestic firms when they both innovate.

Chapter 11

EMPIRICAL LITERATURE ON ENVIRONMENTAL REGULATIONS AND COMPETITIVENESS

In this chapter, we will try to summarize the findings of the empirical literature on the impact of environmental regulations on the international competitiveness of the firms. By using the alternative indicators of competitiveness summarized by Jaffe et al. (1995), mentioned in earlier chapters, we will decompose our analysis into following main sections: Relocation of firms in countries with relatively less stringent environmental regulations, also known as pollution haven hypothesis, the impact of environmental regulations on net exports, and the direction of foreign direct investment as related to the stringency of environmental regulations. We will also mention empirical evidences from firm-level studies and case studies from developing countries.

1. Changes in Trade Patterns–Pollution Haven Hypothesis

The empirical literature that will be summarized here is referred to in the assessment of both the impact of trade liberalization on the level of environmental quality, and the impact of environmental regulations on the international competitiveness. The latter is the natural outcome of such empirical studies, but the former interpretation is necessitated by the lack of reliable cross-country pollution data. Given that the impact of trade liberalization policies on the level of pollution can not be tested in a cross-country study directly, changes in trade patterns as related to the pollution-intensive and non-pollution-intensive industries give useful indirect indications on the national pollution level. Nevertheless, this approach is subject to an important criticism as summarized in Copeland and Taylor (2001):

> An immediate limitation of these studies is apparent: by measuring trends in dirty industry output rather than pollution levels, they have necessarily assumed changes in the composition of a country's output correspond to changes in environmental quality. But if the techniques of production change over time because of trade, income growth

S. Alpay (ed.), Trade and The Environment, 143–156.

Table 11.1. **Ranking of the Dirtiest Manufacturing Industries**

Rank	Air	Water	Metals	Overall
1	Iron and Steel	Iron and Steel	Non-Ferrous Metals	Iron and Steel
2	Non-Ferrous Metals	Non-Ferrous Metals	Iron and Steel	Non-Ferrous Metals
3	Non-Metallic Minerals	Pulp and Paper	Industrial Chemicals	Industrial Chemicals
4	Petro. Coal Prod.	Mis. Man.	Leather Prod.	Petro Refineries
5	Pulp and Paper	Industrial Chemical	Pottery	Non-Metallic Minerals
6	Petro Refineries	Other Chemicals	Metal Products	Pulp and Paper
7	Indust. Chemical	Beverages	Rubber Product	Other Chemicals
8	Other Chemicals	Food Products	Electrical Products	Rubber Products
9	Wood Products	Rubber Products	Machinery	Leather Products
10	Glass Products	Petro. Products	Non-Metallic Minerals	Metal Products

Source: Mani et al. (1997).

or technological progress then a greater share of dirty good output is consistent with both greater and lesser pollution levels. ... At best this method may be able to identify a compositional shift consistent with the pollution haven hypothesis, but never reveal its environmental consequences.

After this small note, we now focus on the regulation-competitiveness linkages. Firstly, a classification of industries with respect to their pollution-intensity (as dirty and clean ones) will be very useful. For such a classification, one can use emission intensity (emissions per dolar of output), toxic intensity (actual discharges per dolar of output) or the percentage of pollution abatement costs in the value-added (see Copeland and Taylor 2001). One such classification is given by Mani et al. (1997) as shown in table 11.1.

Given the decomposition of industries into dirty and clean ones, then one needs to identify the shifts in trade pattern across these industries. Low and Yeats (1992) analyze the world trade pattern in the relevant industries across developing and developed countries for the period 1965-1988. They show that the share of pollution-intensive products in total world trade declined to 16 percent from 19 percent; in North America, a decline of 7 percent was observed (from 21 percent to 14 percent). On the other hand, the share of dirty-good exports in many developing countries rose; for example, in Southeast Asia, it increased to 8.4 percent from 3.4 percent, in Eastern Europe from 21 percent to 28 percent, in Latin America from 17 percent to 21 percent, and in West Asia from 9.2 percent to 13 percent. At the first glance, these results indicate that developing countries became more competitive in pollution-intensive industries, and thus given the differences in the stringency of environmental policies across developed and developing countries one is tempted to conclude that environmental regulations lower the competitiveness of the firms. Similarly, in another

Table 11.2. Largest Exporters of Pollution-Intensive Goods (1988)

Rank	Exporting Country	Value ($ million)	World Trade Share (%)	Share in Country Exports (%)
1	Germany	45.9	11.9	15.8
2	United States	28.5	7.4	10.5
3	Canada	25.2	6.6	23.8
4	France	22	5.7	14.6
5	Belgium-Luxembourg	20.8	5.4	23.5
6	Netherlands	20.3	5.3	20.2
7	Japan	18.9	4.9	8.1
8	United Kingdom	17.3	4.5	14.1
9	Italy	16	4.3	13.8
10	Sweden	15.3	4	33
11	Finland	10	2.6	52.3
12	Soviet Union	8.3	2.2	29
13	Brazil	7.9	2.1	24.3
14	Austria	6.9	1.8	24.6
15	Spain	6.8	1.8	18.4
16	Rep. of Korea	6.6	1.7	11.8
17	Taiwan, China	6.2	1.6	10
18	Norway	6	1.6	27.3
19	Australia	5.7	1.5	19.8
20	Switzerland	5.6	1.5	11.2

Source: Low and Yeats (1992).

cross-country study incorporating 80 countries, Lucas et al. (1992) look at the composition of industrial output and GDP with respect to toxic intensity during the period 1960-1988. They find that the composition of GDP is biased towards cleaner products in rich countries. Moreover, the greatest toxic intensity growth is seen in developing countries. The key message of the paper is that increases in the stringency of environmental regulations in developed countries have led to relocation of pollution-intensive industries to developing countries. Nevertheless, as indicated by Jaffe et al. (1995), industrialized countries accounted for the lion's share of the world's exports of pollution-intensive goods for the period 1965-1988 despite the changes in the shares mentioned above; table 11.2 shows the corresponding data for 1988.

More recently, Mani et al. (1997) study pollution-intensity of output and exports for several developing country regions, Europe, North America, and Japan for the period 1965-1995. Their conclusions are supportive of pollution haven hypothesis: (i) the share of pollution-intensive goods in total manufacturing declined in the OECD countries and increased in developing countries, and (ii) the periods in which the cost of pollution abatement increases in OECD countries coincided with increases in exports of pollution-intensive goods from developing countries. In another study on OECD countries, Xu (1999) finds that the share of environmentally sensitive goods in exports declined to 18 percent in 1995 from 24 percent in 1965. Nevertheless, econometric analysis did not generate evidence for a decline in competitiveness as a result of higher environmental regulations.

Some other papers have studied the competitiveness-regulation interactions in a similar framework for individual countries. For example, Ratnayake (1998) analyzes trade patterns of New Zealand. Imports of pollution-intensive goods from OECD countries dropped to 86 percent in 1993 from 96 percent in 1980. During the same period, the share of pollution-intensive goods imported from the developing countries jumped to 11 percent from 3 percent. On the other hand, exports of pollution-intensive goods from New Zealand to developing countries declined to 46 percent from 59 percent. Below, we will present other case studies on individual countries.

Finally, some other papers approached the competitiveness-regulation issue by studying the impact of *environmental control costs*, henceforth ECC, on industry price and output, and on trade balance. Magee and Ford (1972), Walter (1973), (1982), D'Arge (1974), Richardson and Mutti (1977), OECD (1978), Ugelow (1982), Pasurka (1985), and Robinson (1988). The methodologies adopted in these papers are quite varied; however, they have some common results: estimates of total ECC by industry tend to be very low; abatement costs do not make up a significant portion of industry costs on average; reductions in output as a result of ECC are very small with the exception of some individual sectors.

1.1 A Criticism

Copeland and Taylor (2001) present a very extensive study on the interactions between trade and environment. Two competing hypothesis regarding the source of international trade have been analyzed in depth both theoretically and empirically. The *pollution haven hypothesis* indicates that if environmental policy differences exist across countries then this will generate incentives for countries to trade, and pollution-intensive goods will be imported by countries with relatively more stringent regulations; thus, dirty industries will locate in countries with lenient environmental regulations. The *factor endowment hypothesis* is a well-known hypothesis linking the direction of trade flows with the production intensity and abundant factor of production. In the present context, they assume that capital-intensive sector generates relatively higher emissions, and then with the openness to international markets, capital-abundant countries export pollution-intensive goods and pollution rises in these countries.

Copeland and Taylor (2001) state that while the evidence in empirical studies mentioned above is consistent with the pollution haven hypothesis, they can also be explained by factor endowment hypothesis. They present a simple model, in which economic development in developing countries is based on rapid capital accumulation. Then, it can easily be shown that trade liberalization will result in the expansion of capital-intensive dirty industries, compatible with the empirical evidences presented above. Thus, both the pollution haven and factor endowment hypothesis seem to be possible explanations; however, theory implies that world price of dirty goods will rise if pollution haven hypothesis holds, and the prices will fall if factor endowment hypothesis holds. Copeland and Taylor present the relative prices of the five dirty industries, namely paper and pulp, iron and steel, industrial chemicals, nonferrous metals, nonmetallic mineral products, for the period 1965-2000. The data clearly demonstrates that with the exception of paper and pulp, all of the prices are either constant or declining. Thus, this evidence is not consistent with pollution haven hypothesis. They suggest that "Future analysis must rely more heavily on theory to suggest alternative hypotheses and discipline inferences from the available data. In particular, we would suggest an investigation of the role of the factor accumulation in determining the pollution intensity of national output. The search for pollution havens in the data has obscured the role of capital accumulation and natural resources must play in determining dirty industry migration and trade".

2. Environmental Policy Stringency and Net Exports

Given the limitations of the specific pollution haven hypothesis approach summarized above, a more systematic approach will link changes in trade patterns to (i) classical industry characteristics such as input costs, technology differences and (ii) an indicator of environmental control costs or a measure of environmental policy stringency. Data on environmental regulatory compliance expenditures are in general only available in the developed countries; these resulted in empirical studies focussing on the impact of environmental control costs on trade patterns in the developed countries only. Nevertheless, by introducing qualitative measures of environmental control costs across countries, developing countries were also included in the empirical analysis in some papers. This section will summarize both approaches.

In a Heckscher-Ohlin model of international trade, Kalt (1988) analyzed the impact of environmental compliance costs on net exports. His sample included 78 US industry categories for the period between 1967 and 1977. He finds a statistically insignificant relationship between net exports and environmental compliance costs for the full sample. For the subsample of industries in which national resource industries are excluded, and the sample is restricted to only manufacturing industries, the effect of compliance costs on net exports is negative, and statistically significant. A further experiment with model specification included the removal of chemical industry, a higly pollution-intensive one, from the sample. Surprisingly, the coefficient of environmental compliance costs became even more negative with higher significance. This rather unexpected result was thought to be due to relatively strong net export performance and heavy regulations that the chemical industry was subject to.

Grossman and Krueger (1993) present an empirical analysis of trade patterns between US and Mexico in order to shed some light on the potential environmental impact of the North American Free Trade Agreement (NAFTA). Their model incorporates three different dependent variables (each used separately in regressions): (i) total US imports from Mexico, (ii) imports under the offshore assembly provisions of the US tariff codes, and (iii) the sectoral pattern of maquiladora[1] activity. The set of independent variables include industry factor intensities, tariff rates, and the ratio of pollution abatement costs to total value-added in respective US industries. Although, in two of the six different cross-industry regressions explaining US imports from Mexico, the impact of environmental control costs was negative and significant, overall, they conclude that the effect of environmental costs on the pattern of trade is quantitatively small and statistically insignificant.

A very recent study by van Beers and van den Bergh (1997) applies gravity trade model in a sample from OECD countries for the year 1992. They incorporate an index for the level of environmental regulations across countries. Their contribution is the separation of statistical analysis across resource-based and non-resource based industries. They find that there is no significant relationship between the exports of pollution-intensive goods and environmental regulatory stringency index; however, when the sample is restricted to only non-resource-based (footloose) sectors, the impact of stricter environmental regulations on exports of dirty goods become statistically significant and negative.

[1] Industrial sites in the border areas between US and Mexico with special provisions.

On the other hand, a strong case for the Porter hypothesis is provided by Albrecht (1998). Albrecht considers a model for international CFC-regulation and the export performance of CFC-using industries like refrigerators, freezers, and air conditioning machines; he demonstrates that when regulation is linked to specific products, as in his case, there is clear evidence for the Porter hypothesis.

The most cited work in this literature is by Tobey (1990). Tobey (1990) provides an empirical test of the hypothesis that stringent environmental policy has caused trade patterns to deviate in goods produced by polluting industries in a Heckscher-Ohlin-Vanek model. His model incorporates a qualitative measure of the stringency of national environmental policies developed by Walter and Ugelow (1979). This index is derived from UNCTAD 1976 survey. Given this measure, Tobey tries to identify the impact of the stringency index on the direction of exports across countries.

The data set includes measurements from five different pollution-intensive industries, namely mining, paper and pulp, chemicals, iron and steel and primary non-ferrous metals, in 23 countries (10 developing and 13 developed countries). The dependent variable in the estimated model is the net exports and the independent variables are environmental stringency index, eleven different resource endowments for the year 1975 (such as capital, labor, land, coal, oil, and minerals). His findings support that environmental policy did not impact trade patterns. Tobey states that these empirical results may simply be due to the fact that the magnitude of environmental expenditures are not sufficiently large to cause a significant effect.

Copeland and Taylor (2001) criticize Tobey's econometric approach. Firstly, each of the five cross-country regressions has only 10 degrees of freedom, and therefore, it is not surprising that most of the estimated coefficients are insignificant. Tobey's main conclusion follow from these insignificant coefficients; the limitation imposed by the model and the coverage of the data set undermines the main conclusion. Secondly, the model does not allow for differences across countries, for example in terms of technologies, and therefore treats developed and developing countries identically. On the other hand, introduction of such heterogeneities will make degrees-of-freedom issue more problematic. Thirdly, Tobey uses environmental stringency index values as they are, without converting them into dummy variables; so, his results critically depend on the functional form assumed in his model. There are some nonparametric estimation methods which do not impose any functional form to start with (the functional form is derived from the data); however, such an estimation method has not been employed in Tobey's study. Copeland and Taylor (2001) thus conclude, "Therefore, while Tobey's analysis is perhaps the most cited study arguing against a link between environmental stringency and trade flows, its conclusions rest on tenuous foundations".

3. Foreign Direct Investment

The changes in the location of foreign direct investment (FDI) also gives valuable information on the impact of environmental policies on trade. It is expected that places with lax environmental conditions will attract most of the FDI's in the world. In particular, one would like to see whether differences in environmental control costs (ECC) have led to industrial flight toward LDCs (Less Developed Countries) where environmental regulation is more lenient. Walter (1982) looks at trends in foreign direct investment by firms from Western Europe, Japan, and the United States for the period 1970-1978. Although there exists a large amount of overseas production in pollution-intensive industries, there is little evidence that it has been influenced by differing ECC.

A study by Knodgen (1979) related to West German FDI finds no evidence of widespread relocation of German industries to places with lax environmental standards. Duerksen and Leonard (1980) find out that host countries that received the most overseas investment in pollution-intensive chemicals, paper, metals and petroleum refining were other industrial countries (and not developing countries)[2].

On the other hand, Xing and Kolstad (2002) investigate the relationship between the level of investment by US industries in other countries and the degree of environmental regulation stringency in those countries. They consider foreign direct investment (FDI) of several US industries, representing industries with high pollution control costs (chemicals and primary metals) as well as industries with more modest pollution control costs (electrical and non-electrical machinery, transportation equipment, and food products). In their model, they use aggregate national sulphur dioxide emissions as proxies for the laxity of environmental regulations in a given country. Their statistical results show that the laxity of environmental regulations in a host country is a significant determinant of FDI from the US for heavily polluting industries, and is insignificant for less polluting industries.

[2]These results, however, should be taken with caution because of the many other factors that influence foreign direct investment. It is usually the case that lower environmental standards are in those countries with low degrees of political stability, and less favorite business environment. These factors may be dominant to lenient environmental regulation.

Given data limitations on environmental regulations in developing countries, empirical studies on FDI and environmental-regulations sometimes took an indirect route and investigated the impact of regulations on domestic plant location. Especially US is a very good canditate for such a study given the differences in the degree of environmental regulations across states. Bartik (1988) looks at the relationship between stringency of environmental regulations and plant location choices. He uses state government air and water pollution control expenditures, average costs of compliance and permissible particulate emissions as measures of environmental regulations. The data set used in the empirical analysis comes from the manufacturing branch plants of the Fortune 500 companies for the period 1972-1978. He could not obtain a statistically significant interaction between the plant location decisions and three different indicators of environmental regulations. Bartik (1989) took a slightly different path and investigated the impact of state-level environmental regulations on the start-up rate of small businesses. In his sample of new small businesses from 19 manufacturing industries in 1976-1982 he showed that the level of environmental regulations affected the start-up rate negatively. The impact was statistically significant but its magnitude was very small, if not negligible.

Friedman et. al. (1992) examine the case of new manufacturing branch plant location in the US by foreign multinational companies. The dependent variable in their model is the plant location, and the set of independent variables includes the ratio of pollution abatement capital expenditures to the gross product in a given state. They find that the degree of environmental regulation stringency is not a statistically significant determinant of plant location. In a similar work, Levinson (1992) studies the US manufacturing industries between 1982-1987, and he could not obtain a significant relationship between the level of environmental regulations in a given state and the location of new plants in that state.

4. Firm-Level Evidences

Revisionist school presents empirical support for its claims usually through case studies on individual firms. Subject to new environmental standards, Ciba-Geigy's dyestuff plant in New Jersey reexamined its wastewater streams, which resulted in two changes in the production process that brought annual cost savings of $740,000 (Dorfman, Muir, and Miller 1992). An environmental research organization, INFORM, investigated activities to prevent waste generation at 29 chemical plants. Out of 181 source-reduction activities they investigated, only one resulted in net cost increase; 68 out of 70 cases which had changes in product yield, was found to have yield increases.

Porter and van der Linde (1995) state that environmental regulations can generate innovation offsets through substitution of less costly materials or better utilization of materials in the process. Dorfman et al. (1992) explores the response of Dow Chemical company to new regulations. When federal and state regulations required that Dow Chemical close certain evaporation ponds used for storing and evaporating wastewater resulting from scrubbing hydrochloric gas with caustic soda, Dow Chemical decided to redesign its production process. Looking into the production process, they discovered to scrube the hydrochloric acid with water first and then caustic soda, which led to the elimination of the need for evaporation ponds, a reduction in its use of caustic soda, and capturing a portion of the waste stream for reuse as a raw material in other parts of the plant. The overall cost of the implementation of the redesign was $250,000. After all these changes, caustic waste was reduced by 6,000 tons per year and hydrochloric acid waste by 80 tons per year; the overall savings per year were $2.4 million.

Boroughs and Carpenter (1991) summarize the response of 3M company to new regulations that will force many solvent users in paper, plastic and metal coatings to reduce its solvent emissions 90 percent by 1995. The company responded by avoiding the use of solvents altogether and instituting coating products with safer, water-based solutions. At one 3M plant, Parkinson (1990) states that $60,000 investment for a water-based carrier, used for coating tablets, generated a reduction of 24 tons per year in air emissions and savings of $180,000 in unneeded pollution control equipment and $15,000 in annual solvent purchases.

Revisionists have suggested that induced innovation (with the imposition of stricter regulation) can generate lasting comparative advantage for regulated firms, if other countries eventually apply stricter regulations and if there are strong "first-mover" advantages conferred upon the first firms to enter the markets for control equipment (see, for example, Gardiner 1994).

Finally, as another type of empirical analysis at the firm-level, Levinson (1996) summarizes a number of surveys on the importance of environmental compliance costs in choosing the location of production facilities as reported by managers of the facilities. As seen in table 11.3 (which is also reprinted in Nordstrom and Vaughan 1999), in most cases, favorable environmental conditions (i.e. lower compliance costs) do not seem to be among the top determinants of the location decision with the exception of, probably, high pollution-intensive manufacturing industries.

Table 11.3. Surveys of the Importance of Environmental Regulations to Plant Locations in the United States

Survey	Sample	Result
Epping (1986)	Survey of manufactures (late 1970) that located facilities 1958-1977	"Favorable pollution laws" ranked 43rd to 47th, out of 84 location factors presented.
Schmenner (1982)	*Fortune* 500 branch plants opening 1972-78	Environmental concerns not among the top 6 items mentioned.
Fortune (1977)	*Fortune's* 1977 survey of 1,000 largest U.S. corporations	11% ranked state or local environmental regulations among the top 5 factors.
Wintner (1982)	68 urban manufacturing firms	29 (43%) mentioned environmental and pollution control regulations as a factor in location choice
Stafford (1985)	162 branch plants built in the late 1970s and early 1980s	Environmental regulations were not a major factor, but more important than in the 1970s. When only self-described "less clean" plants were examined, environmental regulations were of "mid-level importance."
Lyne (1990)	*Site Selection* magazine's 1990 survey of corporate real estate executives	Asked to pick 3 of 12 factors affecting location choice, 42% included "state clean air legislation."
Alexander Grant and Company (various years)	Survey of industry associations	Environmental compliance cost given an average of 4%, though growing slightly over time.

Source: Levinson (1996).

5. Individual Country Studies

In this section, we will briefly summarize experiences of several developing countries regarding the impact of environmental regulations on the competitiveness of the firms; these summaries are taken from Jha et al. (1999) mentioned in the preceeding chapters. The edited collection by Jha et al. contains case studies on 11 developing countries (namely Brazil, China, Colombia, Costa Rica, India, Malaysia, Philippines, Poland, Thialand, Turkey, and Zimbabwe). Here, we will mention only the ones with significant changes in their competitiveness that can be associated with environmental regulations.

Brazil managed to keep trade effects of own environmental regulations negligible, by fiscal incentives and special support programmes. Brazil was more vulnerable to external environmental requirements. This was due to (i) the high natural resource and energy intensity of exports, (ii) the high share of exports of homogenous products, with little differentiation potential, competing on the basis of price and (iii) the large share of exports to markets with high environmental standards. A large share of Brazilian exports are destined to OECD countries. Large firms did not have much difficulty in complying with the environmental regulations of these countries, while small and medium firms suffered considerably.

In China, domestic environmental regulations have been introduced in 1970s, and they have helped improve both the environmental and export performances of the firms which have made changes in their operations for better compliance. Foreign regulations caused some loss of export shares; however, again, those firms who have adopted environmentally friendly technologies have generated reduction in wastes and increased their profits.

In Malaysia, domestic regulations reduced timber production but, in the palm oil and electronics sector, export growth was seen during the implementation of domestic environmental regulations. On the other hand, timber, textile, air conditioners and electronics sectors were vulnerable to external regulations. Timber exports and the earnings from exports have fallen drastically due to stringent environmental regulations.

For Philippines' exporters, external environmental regulations did not present significant threats, and the requirements were affordable. The most vulnerable sector to future regulations is sustainable forestry management. Furniture exports will be adversely affected. As regards to the compliance with domestic environmental regulations, the share of abatement cost as a percentage of total sector cost was less than 1% in 51 sectors out of 61 sectors included in the case study. This shows that local environmental standards do not have a significant impact on overall sectoral competitiveness.

In Thailand, exporters were able to adapt themselves to external environmental measures, in general. 42.2% of total exports are estimated to be sensitive to external regulations. Shrimp and shellfish sectors were affected negatively by Japanese Anti-additive Import Regulation of 1991, and secondly by the US restrictions. Thai producers were able to comply with all these restrictions. In textile case, however, small firms faced difficulties when Germany imposed new restrictions in 1995.

6. Summary of Empirical Findings

What can we infer from these empirical studies? Do changes in environmental stringency affect competitiveness adversely or positively? How significant are the results? A recent comprehensive literature survey by Jaffe et. al. (1995) concludes that there is not enough evidence to single out either conventional or revisionist view:

> International differences in environmental regulatory stringency pose insufficient threats to U.S. industrial competitiveness to justify substantial cutbacks in domestic environmental regulations. Nor does the evidence recommend enactment of stricter domestic environmental regulations in order to stimulate economic competitiveness.

At the firm-level, the impact of environmental regulations on the competitiveness would change from sector to sector, and among different firms. The interactions between regulations and competitiveness depend on many firm-specific factors, some of which are listed below (see Jha et al. 1999):

- Destination of exports (whether the firm's export market has stricter or laxer environmental regulations)

- Cost structures (whether the share of environmental compliance costs in the total cost is high, moderate or small)

- Basis for export competitiveness (whether firm exports into a market where competition is largely based on the ability to sell at low prices)

- Firm size (in general, it is relatively more difficult for small firms to comply with environmental regulations than it is for large firms)

- Availability of raw materials, specialized inputs, technology and information (whether firm has easy access to key factors and technologies for the production of environment-friendly products)

- Corporate structures (whether firm has strong vertical integration, as the more a firm is vertically integrated, the greater its ability to comply with environmental regulations)

- Relationship with foreign firms (links with established foreign firms, with greater expertise, make compliance with environmental regulations easier, in general).

Finally, it will be a mistake to see the Porter hypothesis as indicating an automatic increase in the competitiveness of the regulated firms. On the contrary, this view clearly suggests that it is the active involvement of the regulated firms in innovation activities that will improve their competitiveness. Furthermore, one of the important competition strategies today is the differentiation. Both the innovation process and the compliance with higher environmental standards contribute to differentiation. In developed countries, it is not uncommon to see the voluntary adoption of environmental management systems (such as ISO 14000) and the environmentally friendly production certificates. ISO 14000 is being applied by many firms (Boudoropoulos and Arvanitoyannis 2000). In 1999, there were 10439 firms with ISO 14000 certification in the world. The top five countries with respect to number of firms with ISO 14000 certification are as follows: Japan (2124), Germany (1400), England (947), Sweden (645) and Taiwan (492). Among the top twenty, there are 12 countries from European Union. Developing countries fall way behind the developed countries with respect to ISO 14000 certification. Some of the benefits firms would acquire with the application of ISO 14000 environmental management system can be listed as follows (Tansey and Worsley 1995, Goodman 1998):

- Compliance with environmental responsibilities and standards
- Waste reduction
- Better environmental performance
- Improved customer relations
- Reductions in operating costs
- Increases in productivity
- Declining paper work

The cost of adoption of ISO 14000 environmental management system should be assessed against these potential gains; such an approach will make the adoption of environmental management systems more attractive and will also contribute to international competitiveness of the firms, especially when one considers the first-mover advantages associated with better environmental performance in light of the increasing demand for higher environmental quality in developed countries.

III

INTERNATIONAL ENVIRONMENTAL AGREEMENTS AND TRADE

Chapter 12

INTERNATIONAL ENVIRONMENTAL AGREEMENTS AND TRADE

Global environment is subject to two adverse externalities. Transnational environmental degradation, a type of negative externality, imposes certain costs not only on the originating country but also on the other countries (ozone-layer depletion is a good example of this). On the other hand, policies aimed to tackle transnational environmental degradation, involve positive externality from which all countries benefit regardless of their taking action. As a result, countries leave it to others to take pro-environmental actions, and this may result in under-production (if not zero) of environmental good. Thus, as in the public good case, global environmental protection is subject to free-riding. In this chapter, we will present the interactions between trade and environment in the context of international environmental agreements with special attention to international trade regime, trade-environment disputes, and free-riding incentives on global environmental protection.

1. Introduction

Given that not many instruments are available for encouraging global cooperation among countries to solve transfrontier environmental problems, trade policies can be used in motivating countries to contribute to global environment. A well-known standard result in economics is that policy tools should be tailored to the problem at hand (Tinbergen 1952) and thus, trade measures should be used only in response to trade problems; however, in the absence of a powerful intergovernmental organization, transfrontier problems do not seem to fall in the domain of this result, and hence, such function of trade policies should not be ignored. Moreover, trade connections (even without any strategic

S. Alpay (ed.), Trade and The Environment, 159–182.

trade policies, like tariffs and subsidies) among countries can help to decrease free-riding incentives on environmental protection of others and increase the number of countries who contribute to environmental protection.

Earlier, Demsetz (1967) presented some very sharp views on this issue, shared by many others. He states that users of a communally owned resource will fail to come to an agreement on managing the use of the resource even though it is in the interest of all users to cooperate and reduce the amount of the resource they use. This is so because, even if this superior situation is attained, every user will attain even higher returns by free-riding on the cooperative behaviour of the others. The only solution according to Demsetz (1967) is some kind of state intervention. Taking these views of Demsetz (1967) as his starting point, Barrett (1990) states that the theoretical arguments for supposing that cooperation will not develop are compelling, but they can not be complete as cooperation (in some form) does take place and is codified in international agreements (whether effective or not). He then goes on to explore the reasons why international cooperation might develop.

Barrett (1994a), and Heal (1994) study international agreements to protect the environment and free-riding[1] in the context mentioned above. Barrett (1994a) shows that self-enforcing international environmental agreements (IEAs) may not be able to improve substantially over the non-cooperative outcome. There are cases when self-enforcing IEAs can sustain a large number of countries, but this requires that the difference in net benefits between the non-cooperative and full-cooperative outcome is very small. Heal (1994) studies the formation of emissions abatement coalitions. In a game theoretical model, Heal shows that the selection of positive abatement levels by other countries may shift a particular country's best choice of abatement from zero to a positive number if the benefit and cost functions of identical countries are not separable across abatement levels ('reinforcement effects'), or if there are fixed costs associated with abatement programs; thus, free-riding need not always be the best action.

These papers, however, do not include connections among countries through international trade of goods and services. A study by Blackhurst and Subramanian (1992) on multilateral cooperation on environmental issues, sets out the obstacles in the path to cooperation (free-riding being the main component), and

[1]In an earlier literature, there has been put some doubt on the importance and the existence of the free rider problem. Explicitly, we can cite Bohm (1972), Sweeney (1973), and Smith (1979), who designed experiments which resulted in negligible free-riding when the agents had strong incentives to free ride. Nevertheless, Kim and Walker (1984) suggested that there were some 'invalidating factors' in these previous experiments and by removing these factors in their experiment, they found out that, in fact, free-riding was substantial.

then they state that trade policies generate incentives for countries to participate in multilateral efforts to deal with environmental problems.

The most efficient way to handle global environmental problems is international cooperation. The decisions pertaining to participation or not to any international initiative on environmental problems depend on many factors other than free-riding on the benefit side. Factors like production substitution across countries, terms of trade changes arising from the international connections among countries play significant roles in the final decision; acting only free-riding incentives will be a short-sighted move. Piggott, Whalley, and Wigle (1992) present empirical evidences for these arguments in their study on carbon emission reductions.

International cooperation on global environmental protection through international environmental agreements (IEA) is a very difficult process to realize, but what is equally difficult is their implementation. In the absence of a supranational authority, IEAs are subject to free-riding, and thus, their chance of being successful is very limited if a proper enforcement mechanism is not developed. An interesting suggestion is the use of international trade as an enforcement tool in the implementation of international environmental agreements (IEA). Barrett (1999) considers the applicability of trade sanctions as an instrument to enforce international environmental agreements aiming to provide a global public good, such as reduction in different pollutants. He shows that trade policies can be used to enforce international environmental agreements provided that (1) every country must be better off as a signatory than as non-signatory when sanctions are imposed against free riders, and (2) when free-riding occurs, signatories must be better-off by imposing sanctions than by continuing to trade with non-signatories.

Alpay (2000) takes up the question whether the suboptimality of the production of a public good in a closed economy extends to open economy case, more explicitly, whether global environmental protection is also subject to free-riding in a model where trade connections among countries are taken into consideration. It has been shown that countries are not always reluctant to contribute to global environmental protection. Even if there is no self-financed transfers between countries, when the terms of trade changes associated with environmental protection are taken into consideration, countries may choose to contribute to global protection instead of free-riding on others' contribution. This non-cooperative contribution, contrary to the conventional results, exceeds that of the cooperative one. As a policy implication, the paper concludes that the assessment of government policies on global environmental protection in a partial equilibrium framework by ignoring the possible trade interactions, may very well be misleading. This paper will be presented in more detail in the next chapter.

A second instrument regarding the implementation of international environmental agreements is income transfer. Carraro and Siniscalco (1993) analyze the profitability and stability (no free-riding) of international environmental agreements in the presence of transboundary pollution. They show that the agreements tend to be realized among a fraction of negotiating countries, and that the number of signatories can be increased through self-financed transfers; however, any expansion of a coalition requires some form of commitment, which is a strong assumption.

A few other studies carry out the analysis in an incomplete information framework. Petrakis and Xepapadeas (1996) analyze global or partial stable coalitions among two groups of countries, which differ from each other in terms of environmental consciousness, through self-financed transfers. They demonstrate a mechanism which detects free-riding (or cheating) even when the individual emissions can not be observed by other countries.

As a real life application of the use of income transfers and trade policies as enforcement tools for an international environmental agreement, we can mention the Montreal Protocol on Substances that Deplete the Ozone Layer. This is one of the most successful international agreements on global environmental protection. The use of trade tools helped reduce and coordinate the transactions of ozone depleting substances among the signatories and non-signatories. Income transfers were affective in attracting developing countries to sign the agreement. More detailed information regarding this agreement will be provided below.

Given that trade policies may be effective in the implementation of international environmental agreements, the interesting question becomes what kind of policies may be acceptable under the current trading rules agreed upon by countries. Given the attempts to liberalize trade starting from the second part of the last century, can countries apply restrictive trade policies for environmental protection? How can countries distinguish genuine trade policies that aim to serve an environmental purpose from the ones that aim to achieve disguised trade protection? To understand these issues better, we now provide a brief introduction on the current international trade regime, its provisions on environmental issues, the dispute handling mechanisms, regional trade blocks and environmental concerns, and finally some of the international environmental agreements.

2. International Trading Regime

In this section, we will see some legal aspects of trade and environment. We will start with the provisons contained in the GATT for environmental protection; then the process of handling disputes among countries in this regard will be presented with some cases from recent past. International environmental treaties with trade instruments will be another subject of this section with emphasis on the Montreal Protocol.

2.1 Trade and the Environment in the GATT

For the last 50 years, the international trade regime has been guided by the GATT (General Agreement on Tariffs and Trade). It has been established with the purpose of promoting a liberal trade environment by minimizing restrictions on trade. In this regard, it has been very successful as can be seen from the expansion of volume of trade in the last part of the previous century.

Main principles forming the skeleton of GATT articles are (i) most-favored nation treatment, (ii) national treatment, (iii) tarrification, and (iv) consultation[2]. Since global environmental problems were not a big issue when GATT was initiated, there is no direct provision for environment in the agreement; however, Article XX lists some exceptions to the general rules, which may be construed to include some environmental measures. Let us mention the relevant contents of this article:

> Subject to the requirement that such measures are not applied in a manner which would constitute a means of arbitrary or unjustifiable discrimination between countries where the same conditions prevail, or a disguised restriction on international trade, nothing in this Agreement shall be construed to prevent the adoption or enforcement by any contracting party of measures:
>
> (b) necessary to protect human, animal or plant life or health; and
> (g) relating to the conservation of exhaustible natural resources if such measures are made effective in conjunction with restrictions on domestic production and consumption.

Uruguay Round of negotiations of the GATT has been the first round where issues on environment have been important items of the agenda and been debated strongly. Goals of the round, such as reducing and eliminating non-tariff barriers including environmental measures, liberalizing trade in natural-resource based

[2]First principle requires that like products from different countries must be treated equally; second one applies the same rule to imported products versus domestic products; third one mandates that all non-tariff barriers, quotas, bans, etc., be converted to tariffs; and the last principle requires parties to try to settle their trade disputes through consultation and negotiation. See, for example, Esty (1994).

products, harmonization of environmental and health standards, are visioned by environmental advocates, as sacrifice of natural resource protection and stimulation of resource depletion. On the pro-environment side, participants agreed on forming Environmental Committee under World Trade Organization (WTO), which was also established in the same round as an international institution to support the GATT. In the appendix, we briefly summarize the establishment of Committee on Trade and Environment (CTE) under WTO.

Trade and environment disputes among members have been addressed by three-person panels of impartial trade experts appointed by director general of the GATT. Recommendations of panels have to be approved by the GATT Council, where each member has a veto power. Moreover, there is no legal enforcement of the approved decisions. There are 8 disputes regarding trade and environment handled by GATT Council since 1947. In all these disputes, complaints are based on the violation of articles XX-b and XX-g, which provide provisions for environmental protection. A complete list and details of these disputes are given in the appendix. We will briefly summarize two of them here.

2.1.1 Canada–US Tuna Case (1982)

In a dispute between US and Canada relating to jurisdiction over Pacific fisheries, 19 US fishing vessels were seized by Canada while operating within Canadian Exclusive Economic Zone, which was responded with a total ban on Canadian tuna and tuna products by the United States. Canada challenged the ban and took the case to the GATT dispute panel. The United States claimed that the ban was justified under GATT Article XX (g) in order to conserve an exhaustible natural resource.

The GATT panel decided that the US measure was not "primarily aimed at" environmental protection as required for an Article XX (g) exception. This decision was based on the fact that (i) US ban did not make any exception for those species of tuna not in danger of depletion and (ii) the United States did not take any steps to restrict domestic tuna consumption accordingly.

2.1.2 Tuna-Dolphin I-(1991) and II-(1994)

The US Marine Mammal Protection Act (MMPA) requires that importation of tuna products from countries whose dolphin kill ratio (per net dropped to catch tuna) exceeds that of US ratio plus a margin, be prohibited. In 1988, Earth Island Institute, an environmental group, sued the Mexican fishermen, importing tuna to the USA in violation of this law. A federal judge agreed that the law was being violated and banned the Mexican tuna imports. A GATT panel was formed at the request of Mexico who argued that GATT rules were violated by US with the restriction of its right to sell tuna in the USA.

The panel found against the US. According to panel, the US ban violated national treatment requirement in Article III; the standard, i.e. the US kill ratio, was deemed inappropriate because it was retroactively determined based on the number of dolphins killed by American fishermen in any year– and so uncertain until after the fact; and the ban could neither be justified under Article XX (b) and (g), and nor these exceptions (exceptions b and g outlined above) could be applied unilaterally or extrajurisdictionally.

Nevertheless, the decision of the panel was not approved by the GATT Council. In 1994, this time the European Community took the case to the GATT, complaining, in general, about the MMPA itself, and in particular, about the secondary embargo provision which bars tuna imports to the USA from any country engaging in tuna trade with an embargoed country. Since several EC countries were importing tuna from Mexico, the secondary embargo became relevant to them. The panel again decided against the ban based on unilateral nature of the US trade restriction. The weakness of the dispute handling mechanism was seen here as well as the final panel decision was not approved by the GATT Council.

We can infer from these cases that GATT is trying to realize a world of free trade, and its objectives do not necessarily include environmental concerns. Of all the eight trade-environment disputes handled by the panels, none is accepted to conform to the terms indicated in the articles XX-b and g. Obviously, countries will have all the incentives to use environmental protection as disguised trade barriers; however, these past experiences have also indications for the bias in GATT panels towards frree trade. Nevertheless, this is not true for all trade–environment disputes. In the Danish Bottles case, the Court of the European Communities justified the decision of Denmark that the drinks could be marketed only in returnable containers, on the grounds of environmental protection, which was taken to the Court by the other EC members on the grounds that Denmark's decision was against its free trade duties under the Single European Act. This judgment, the first that restricts free trade on environmental grounds, rested on the argument that Denmark's case provided good grounds for finding that protection of the environment constitutes an imperative requirement which may limit the application of the free trade rules (for a detailed exposition, see, for example, Clark and Arnold 1993). Given the experiences in the past, there are proposals regarding the establishment of a new and separate international organization on environment, named by many as World Environment Organization. As mentioned earlier an alternative route is the transformation of GATT into GATE (General Agreement on Trade and the Environment). On these issues one can see Runge (1994, ch.6), Rauscher (1997, ch. 9) and Cole (1999, ch. 7).

Now, we would like to present the role of environmental issues in the establishment of a regional trade agreement: NAFTA. Given the public pressure on environmental issues, the treatment of environment under NAFTA has been done much more satisfactorily as compared to the GATT. We now expose some of the details regarding environmental provisions in NAFTA.

2.2 Trade and the Environment in NAFTA

North American Free Trade Agreement (NAFTA) establishes the largest regional trade bloc in the world, with $6.5 trillion in GDP, $243 billion in intrabloc trade, $760 billion in extrabloc trade, and 360 million consumers. First draft came out in 1992, and the North American Free Trade Agreement took effect on January 1, 1994, among Canada, Mexico, and the United States. In the six years since NAFTA was implemented, total trade in North America is estimated to have grown to approximately US$700 billion per year.

Environmental concerns are more pronounced in NAFTA as compared to GATT. This can be seen in the following articles of NAFTA regarding the environmental issues[3]:

> In the event of any inconsistency between this Agreement and the specific trade obligations set out in:
>
> a) the Convention on International Trade in Endangered Species of ...
>
> b) the Montreal Protocol on Substances that Deplete the Ozone Layer ...
>
> c) the Basel Convention on the Control of Transboundary Movements of Hazardous Wastes and Their Disposal ...
>
> d) The Agreement Between the Government of Canada and the Government of the United States of America Concerning the Transboundary Movement of Hazardous Waste ...
>
> The Agreement Between the United States of America and the United Mexican States on Cooperation for the Protection and Improvement of the Environment in the Border Area ...
>
> such obligations shall prevail to the extent of the inconsistency, provided that where a Party has a choice among equally effective and reasonably available means of complying

[3]For full text of NAFTA, one can see http://www.dfait-maeci.gc.ca/nafta-alena/agree-e.asp.

> with such obligations, the Party chooses the alternative that is the least inconsistent with the other provisions of this Agreement. (Article 104)
>
> Each Party may, in accordance with this Agreement, adopt, maintain or apply any standards-related measure, including any such measure relating to safety, the protection of human, animal or plant life or health, the environment or consumers, and any measure to ensure its enforcement or implementation. Such measures include those to prohibit the importation of a good of another Party or the provision of a service by a service provider of another Party that fails to comply with the applicable requirements of those measures or to complete the Party's approval procedures. (Article 904)
>
> The Parties recognize that it is inappropriate to encourage investment by relaxing domestic health, safety or environmental measures. Accordingly, a Party should not waive or otherwise derogate from, or offer to waive or otherwise derogate from, such measures as an encouragement for the establishment, acquisition, expansion or retention in its territory of an investment of an investor. If a Party considers that another Party has offered such an encouragement, it may request consultations with the other Party and the two Parties shall consult with a view to avoiding any such encouragement. (Article 1114)

For enhanced environmental protection and conservation in the long-run, the U.S. and Mexican governments are pursuing under NAFTA a series of long-term cooperative programs covering pollution control and prevention, pesticides, waste management, and emergency response, among other things. This work is expected to result in the establishment of new mechanisms for environmental cooperation, including enforcement.

Environmental issues among the member countries are being handled by the Commission for Environmental Cooperation (CEC). CEC addresses regional environmental concerns in North America, helps prevent potential trade and environmental conflicts, and promotes the effective enforcement of environmental law, all as part of its mandate under the North American Agreement on Environmental Cooperation (NAAEC).

There are very differing views on the success of NAFTA on environmental protection. One view asserts that the increase in the income of Mexico through free trade would provide funds for environmental protection, and rise environmental standards. A study by Grossman and Krueger (1991) conclude that NAFTA is actually likely to improve environmental quality in Mexico, in support of this view.

A second view challenged this idea and argued that NAFTA would essentially promote industrial growth, and in the absence of attention to environmental provisions in the agreement, this would worsen the environmental conditions, especially in Mexico.

A third view supports sustainable development, and proposes that both free trade and environmental protection are valuable. Supporters of this view argue that while trade will open channels towards affluence and generate funding for environmental programs, it must comply with the environmental objectives to avoid further environmental degradation; however, the weights of free trade and environmental considerations in sustainable development, is not easy to assess. NAFTA's balance on this issue is a matter of dispute.

In the next section, we present a summary of recent findings regarding the environmental impacts of NAFTA.

2.3 Environmental Impact of NAFTA

This section is based on "The Environmental Effects of Free Trade[4]," which is the edited collection of papers presented at the North American Symposium on Assessing the Linkages between Trade and Environment, October 2000. The following passage is a brief summary of the remarks by P. M. Johnson, the chairman of the symposium, on the presented papers.

The symposium papers suggest that the environmental effect of NAFTA is mixed. For example, in the fisheries sector, no significant impact has been identified. On the other hand, for forest products, the restructuring of the industry has been accompanied by significant changes, including its exposure to contestation from international competition. Regarding the particular pollutants, some air pollution indicators show increases of carbon monoxide and sulfur dioxide levels in the US, and of sulfur dioxide in Mexico, and significant reductions of air pollution in the Canadian and Mexican paper sectors. Hazardous waste generation in some Canadian provinces is increasing along with a decline in some northern US states; total amount of trade in hazardous wastes, in particular, waste imported into Canada from the United States, has increased significantly after NAFTA. One of the strongest links between international trade and environmental effects is in the transport sector. The data show that, while there has been an absolute increase in truck transport through concentrated border crossings, these growth rate increases are not related to rates of environmental change.

With respect to the impact of trade liberalization on the stringency of environmental regulations (in particular "race to the bottom" hypothesis) there is no empirical support for NAFTA's leading to the weakening of environmental

[4]This book is available at http://www.cec.org/pubs_docs/scope/index.cfm?varlan=english&ID=14.

regulations.

The impact of NAFTA on Mexico's environment is one of the most debated issue, if not the top. Has Mexico turned into a "pollution haven"? The evidence suggests that Mexican export specialization has resulted in less and less pollution. In contrast, Canadian export specialization is now much more pollution-intensive than Mexico's. Among the interesting papers of the symposium are those that examine the services sector. While many think that the services sector is by definition "clean," in fact it is becoming an increasingly important source of pollution. New approaches to assessing the services sector in terms of its pollution by-product are needed.

2.4 International Environmental Agreements

In this part, we will focus on international environmental treaties. Global environmental problems do not recognize national frontiers and sovereignty[5], so the only hope to solve these problems is to initiate and mobilize international cooperation through binding treaties. Contrary to the expectations, there are quite a number of multilateral/bilateral treaties on the environmental issues; for example, Register of International Treaties and Other Agreements on the Field of Environment identifies 152 multilateral agreements. Nevertheless, it is needless to say that most of these agreements do not bring desired outcomes due to implementation problems.

International cooperation on environmental issues gained momentum after the Stockholm Conference in 1972. At this conference, countries agreed to establish the United Nations Environment Programme (UNEP) to incorporate environmental considerations into the operations of other UN organizations, like FAO, UNESCO, and to adopt a number of recommendations on international cooperation. As one of the important decisions of the Stockholm Declaration, paragraph 21 states that: "States have ... responsibility to ensure that activities within their jurisdiction or control do not cause damage to the environment of other States or of other areas beyond the limits of national jurisdiction." For more information on this declaration, and an introduction to international environmental problems, Mäler (1990) should be seen.

An effective treaty, as in the case of Montreal Protocol on Substances That Deplete the Ozone Layer[6], requires commitment, consensus and enforcement.

---

[5] At least those with transnational feature.
[6] More on this protocol below.

Commitment depends on how important the treaty is to each participant, which in turn depends on benefit-cost analysis. Consensus on the nature of the problem and on the possible solution shapes the treaty itself. Finally, the treaty must include enforcement tools. In the absence of an international agency with ability to monitor and enforce the environmental treaties, the role of other enforcement tools is very vital. Two possible tools are subsidies (or compensation funds) and trade sanctions (or trade policies). Subsidies usually require willingness on the rich participants of the treaties, and trade sanctions may require compliance to the GATT.

Trade sanctions seem to work fine as an enforcement tool; it has been successfully used in at least two agreements: Montreal Protocol and Convention on International Trade in Endangered Species. Now, we continue with a detailed exposition of the process that led to reduction of CFCs successfully: the Montreal Protocol.

2.4.1 Montreal Protocol on Substances That Deplete the Ozone Layer

The ozone layer surrounds the earth's upper atmosphere and holds the sun's ultraviolet rays, which have detrimental health effects like skin cancer, together with other biological and environmental problems. Early 1970s, scientists expressed their concerns about the increased concentrations of chloroflurocarbons (CFC) which would cause depletion of ozone layer; however, an established link between CFC use and ozone-depletion could not be shown. In 1977, the US unilaterally banned CFC's as aerosol preplans. Shortly after, Canada, Sweden, Norway, Denmark took similar actions.

In 1981, a legal and technical working group was formed by UNEP to draft a Global Framework Convention for the Protection of the Ozone Layer. This led to the Vienna Convention for the Protection of the Ozone Layer with the participation of 43 countries in March 1985. The Vienna Convention called for international cooperation in research, monitoring, and information exchange; in this convention a "concrete" agreement on the control of CFCs could not be reached but it set a framework for a more effective treaty on the issue.

In 1985, a British research disclosed a huge, seasonal ozone hole over Antarctica; there were also evidences of depletion over the mid-latitudes in a smaller scale. In September 1987, UN efforts since 1981 gave fruits and 24 countries signed the Montreal Protocol to protect the ozone layer. The treaty (i) included 50 percent reduction in CFC production from a 1986 baseline and a freeze in halon production; (ii) developing countries were given a ten-year grace period before any cutbacks; (iii) an assistance fund was to be set up by rich countries to ease the transition of developing countries (however, this was not required); and (iv) as the enforcement policy to discourage free-riding by non-members, parties to the treaty were not allowed to export or import CFCs, many products containing and produced with CFCs, and CFC technologies to nonmember countries[7]. Under these sanctions, free-riding was not easy and so, more countries decided to sign the agreement. Trade provisions listed under article 4 of the Montreal Protocol are as follows:

Article 4: Control of Trade with Non-Parties

1. Within one year of the entry into force of this Protocol, each Party shall ban the import of controlled substances from any State not party to this Protocol.

2. Beginning on 1 January 1993, no Party operating under paragraph 1 of Article 5 may export any controlled substance to any State not party to this Protocol.

3. Within three years of the date of the entry into force of this Protocol, the Parties shall, following the procedures in Article 10 of the Convention, elaborate in an annex a list of products containing controlled substances. Parties that have not objected to the annex in accordance with those procedures shall ban, within one year of the annex having become effective, the import of those products from any State not party to this Protocol.

4. Within five years of the entry into force of this Protocol, the Parties shall determine the feasibility of banning or restricting, from States not party to this Protocol, the import of products produced with, but not containing, controlled substances. If determined feasible, the Parties shall, following the procedures in Article 10 of the Convention, elaborate in an annex a list of such products. Parties that have not objected to it in accordance with those procedures shall ban or restrict, within one year of the annex having become effective, the import of those products from any State not party to this Protocol.

5. Each Party shall discourage the export, to any State not party to this Protocol, of technology for producing and for utilizing controlled substances.

6. Each Party shall refrain from providing new subsidies, aid, credits, guarantees or insurance programmes for the export to States not party to this Protocol

[7]Note that this enforcement policy through trade barriers could be challenged under GATT regulations by non-member countries; however, most members of the GATT are also members of the Montreal Protocol, so such challenges need not diminish the effectiveness of this enforcement policy. Trade works for environment here.

of products, equipment, plants or technology that would facilitate the production of controlled substances.

7. Paragraphs 5 and 6 shall not apply to products, equipment, plants or technology that improve the containment, recovery, recycling or destruction of controlled substances, promote the development of alternative substances, or otherwise contribute to the reduction of emissions of controlled substances.

The treaty entered into force on January 1, 1989. After the new scientific findings on the seriousness of the depletion and the link between CFC and depletion, in 1990, a full phaseout of CFCs, halons, and other ozone-depleting materials by the year 2000 was amended to the protocol. In 1992, based on a finding that an accelerated phaseout might save an additional million of lives through reduced skin cancer, the participants agreed to eliminate CFCs by 1996. Also, note that when the initial incentives were not enough to attract India and China, a 1990 revision of the treaty required rich participants to establish a fund of $260 million to finance the transition period in low income participants; this fund was raised to $500 million after a 1992 revision[8].

[8] See Goodstein (1995).

3. Appendix

The following passages on CTE are taken from WTO documents with minor modifications[9]. They have been reprinted in Nordstrom and Vaughan (1999).

3.1 The Committee on Trade and Environment (CTE)

Towards the end of the Uruguay Round, GATT contracting parties agreed that the Trade Negotiations Committee (TNC) should adopt a work programme on trade and environment and present it, together with recommendations on an institutional structure for its execution, at the Marrakesh Ministerial Conference. This led to the adoption, on 14 April 1994, of the Decision on Trade and Environment (hereinafter the "Marrakesh Decision") in which Trade Ministers noted that it should not be contradictory to safeguard the multilateral trading system on the one hand, and act for the protection of the environment and the promotion of sustainable development on the other hand. Ministers further noted their desire to coordinate policies in the field of trade and environment, "but without exceeding the competence of the multilateral trading system, which is limited to trade policies and those trade-related aspects of environmental policies which may result in significant trade effects".

The Marrakesh Decision directed the first meeting of the General Council of the WTO to establish a Committee on Trade and Environment (CTE), whose tasks are: "to identify the relationship between trade measures and environmental measures, in order to promote sustainable development; (b) to make appropriate recommendations on whether any modifications of the provisions of the multilateral trading system are required, compatible with the open, equitable and non-discriminatory nature of the system ...". The Marrakesh Decision lists ten items, encompassing all areas of the multilateral trading system: goods, services and intellectual property. These items are commonly referred to in the following order:

Item 1: "the relationship between the provisions of the multilateral trading system and trade measures for environmental purposes, including those pursuant to multilateral environmental agreements"

Item 2: "the relationship between environmental policies relevant to trade and environmental measures with significant trade effects and the provisions of

[9] See the web site http://www.wto.org/english/tratop_e/envir_e/tr_envbadoc.htm.

the multilateral trading system"

Item 3: "the relationship between the provisions of the multilateral trading system and:

(a) charges and taxes for environmental purposes

(b) requirements for environmental purposes relating to products, including standards and technical regulations, packaging, labelling and recycling"

Item 4: "the provisions of the multilateral trading system with respect to the transparency of trade measures used for environmental purposes and environmental measures and requirements which have significant trade effects"

Item 5: "the relationship between the dispute settlement mechanisms in the multilateral trading system and those found in multilateral environmental agreements"

Item 6: "the effect of environmental measures on market access, especially in relation to developing countries, in particular to the least developed among them, and environmental benefits of removing trade restrictions and distortions"

Item 7: "the issue of exports of domestically prohibited goods"

Item 8: "The relevant provisions of the Agreement on Trade-related Aspects of Intellectual Property Rights"

Item 9: "The work programme envisaged in the Decision on Trade in Services and the Environment"

Item 10: "appropriate arrangements for relations with non-governmental organizations referred to in Article V of the WTO and transparency of documentation."

As stipulated in the Marrakesh Ministerial Decision on Trade and Environment, the General Council of the WTO established the Committee on Trade and Environment (CTE) at its first meeting, held on 31 January 1995. The CTE held its first meeting on 16 February 1995. It adopted a programme of work whereby each meeting would focus on some of the ten agenda items. CTE Members also agreed that meetings would be organized such that, once discussion of the items constituting the focus of the meeting had been completed, delegations could address, if they wished, the item(s) that had been discussed at the previous meeting. The work of the CTE was assisted by background and analytical papers prepared by the Secretariat, as well as documents submitted by delegations.

The CTE has currently granted observer status to twenty intergovernmental organizations: the United Nations (UN), the United Nations Conference on Trade and Development (UNCTAD), the World Bank, the International Monetary Fund (IMF), the United Nations Environment Programme (UNEP), United Nations Development Programme (UNDP), Commission for Sustainable Development (CSD), the Food and Agriculture Organization (FAO), the International Trade Centre (ITC), the Organization for Economic Cooperation and Development (OECD), and the European Free Trade Association (EFTA), African, Caribbean and Pacific Group of States (ACP Group), Convention on Biological Diversity (CBD), Convention on International Trade in Endangered Species of Wild Fauna and Flora (CITES), International Organization for Standardization (ISO), International Plant Genetic Resources Institute (IPGRI), Latin American Economic System (SELA), United Nations Industrial Development Organization (UNIDO), World Customs Organization (WCO), World Intellectual Property Organization (WIPO).

3.2 Trade and Environment Disputes

From 1947 to 1995, of the 115 panel reports issued under the GATT 1947 and the Tokyo Round agreements, 6 concerned issues related to human and animal health or the environment. Since the entry into force of the WTO, on 1 January 1995, 6 such cases can be identified among the 38 panels which have been established so far. This section presents a brief summary of these cases[10].

United States — Prohibition of Imports of Tuna and Tuna Products from Canada, adopted on 22 February 1982.

An import prohibition was introduced by the US after Canada had seized 19 fishing vessels and arrested US fishermen fishing for albacore tuna, without authorization from the Canadian government, in waters considered by Canada to be under its jurisdiction. The US did not recognize this jurisdiction and introduced an import prohibition to retaliate under the Fishery Conservation and Management Act.

---

[10] We will only summarize the eight out of the twelve cases mentioned. The distinction is based on whether the disputes have been based on articles XX-b and XX-g, which provide explicit provisions for environmental protection. Following passages are taken from http://www.wto.org/english/tratop_e/envir_e/tr_envbadoc.htm.

The Panel found that the import prohibition was contrary to Article XI: 1, and was justified neither under Article XI: 2, nor under Article XX (g) of the General Agreement.

Canada — Measures Affecting Exports of Unprocessed Herring and Salmon, adopted on 22 March 1988

Under the 1976 Canadian Fisheries Act, Canada maintained regulations prohibiting the exportation or sale for export of certain unprocessed herring and salmon. The US complained that these measures were inconsistent with GATT Article XI. Canada argued that these export restrictions were part of a system of fishery resource management aimed at preserving fish stocks, and therefore were justified under Article XX (g).

The Panel found that the measures maintained by Canada were contrary to GATT Article XI: 1 and were justified neither by Article XI: 2(b) nor by Article XX (g).

Thailand — Restrictions on the Importation of and Internal Taxes on Cigarettes, adopted on 7 November 1990

Under the 1966 Tobacco Act, Thailand prohibited the importation of cigarettes and other tobacco preparations, but authorized the sale of domestic cigarettes; moreover, cigarettes were subject to an excise tax, a business tax and a municipal tax. The US complained that the import restrictions were inconsistent with GATT Article XI: 1, and considered that they were justified neither by Article XI: 2(c), nor by Article XX (b). It also argued that the internal taxes were inconsistent with GATT Article III: 2. Thailand argued, inter alia, that the import restrictions were justified under Article XX (b) because the government had adopted measures that could only be effective if cigarette imports were prohibited and because chemicals and other additives contained in US cigarettes might make them more harmful than Thai cigarettes.

The Panel found that the import restrictions were inconsistent with Article XI: 1 and not justified under Article X1:2(c). It further concluded that the import restrictions were not "necessary" within the meaning of Article XX (b). The internal taxes were found to be consistent with Article III: 2.

United States — Restrictions on Imports of Tuna, not adopted, circulated on 3 September 1991

The Marine Mammal Protection Act (MMPA) prohibited the "taking" (harassment, hunting, capture, killing or attempt to do any of these), and importation into the US, of marine mammals, except with explicit authorization. It governed, in particular, the taking of marine mammals incidental to harvesting yellowfin tuna in the Eastern Tropical Pacific Ocean, an area where dolphins are known to swim above schools of tuna. Under the MMPA the importation of commercial fish or products from fish which have been caught with commercial fishing technology which results in the incidental kill or incidental serious injury of ocean mammals in excess of US standards were prohibited. In particular, the importation of yellowfin tuna harvested with purse-seine nets in the Eastern Tropical Pacific Ocean was prohibited (*primary nation embargo*), unless the competent US authorities established that:

(i) The government of the harvesting country had a programme regulating the taking of marine mammals, comparable to that of the US, and

(ii) The average rate of incidental taking of marine mammals by vessels of the harvesting nation was comparable to the average rate of such taking by US vessels.

The average incidental taking rate (in terms of dolphins killed each time in the purse-seine nets are set) for that country's tuna fleet were not to exceed 1.25 times the average taking rate of US vessels in the same period. Imports of tuna from countries purchasing tuna from a country subject to the primary nation embargo were also prohibited *(intermediary nation embargo).*

Mexico claimed that the import prohibition on yellowfin tuna and tuna products was inconsistent with Articles XI, XIII and III of GATT . The US requested the Panel to find that the direct embargo was consistent with Article III and, in the alternative, was covered by Articles XX (b) and XX (g). The US also argued that the intermediary nation embargo was consistent with Article III and, in the alternative, was justified by Article XX, paragraphs (b), (d) and (g).

The Panel found that the import prohibition under the direct and the intermediary embargoes did not constitute internal regulations within the meaning of Article III, was inconsistent with Article XI:1 and was not justified by Article XX paragraphs (b) and (g). Moreover, the intermediary embargo was not justified under Article XX (d).

United States — Restrictions on Imports of Tuna, not adopted, circulated on 16 June 1994

The European Communities (EC) and the Netherlands complained that both the *primary* and the *intermediary* nation embargoes, enforced pursuant to the Marine Mammal Protection Act, did not fall under Article III, were inconsistent with Article XI:1 and were not covered by any of the exceptions of Article XX. The US considered that the *intermediary* nation embargo was consistent with GATT since it was covered by Article XX, paragraphs (g), (b) and (d), and that the *primary* nation embargo did not nullify or impair any benefits accruing to the EC or the Netherlands since it did not apply to these countries.

The Panel found that neither the *primary* nor the *intermediary* nation embargo were covered under Article III, that both were contrary to Article XI:1 and not covered by the exceptions in Article XX (b), (g) or (d) of the GATT.

United States — Taxes on automobiles, not adopted, circulated on 11 October 1994

Three US measures on automobiles were under examination: the luxury tax on automobiles ("luxury tax"), the gas guzzler tax on automobiles ("gas guzzler"), and the Corporate Average Fuel Economy regulation ("CAFE"). The European Community complained that these measures were inconsistent with GATT Article III and could not be justified under Article XX (g) or (d). The US considered that these measures were consistent with the General Agreement.

The Panel found that both the luxury tax (which applied to cars sold for over $30,000) and the gas guzzler tax (which applied to the sale of automobiles attaining less than 22.5 miles per gallon (mpg)) were consistent with Article III: 2 of GATT.

The CAFE regulation required the average fuel economy for passenger cars manufactured in the US or sold by any importer not to fall below 27.5 mpg. Companies that were both importers and domestic manufacturers had to calculate average fuel economy separately for imported passenger automobiles and for those manufactured domestically.

The Panel found the CAFE regulation to be inconsistent with GATT Article III: 4 because the separate foreign fleet accounting system discriminated against foreign cars, and the fleet averaging differentiated between imported and domestic cars on the basis of factors relating to control or ownership of producers or importers, rather than on the basis of factors directly related to the products as such. Similarly, the Panel found that the separate foreign fleet accounting was not justified under Article XX (g); it did not make a finding on the consistency of the fleet averaging method with Article XX (g). The Panel found that the CAFE regulation could not be justified under Article XX (d).

United States — Standards for Reformulated and Conventional Gasoline, adopted on 20 May 1996

Following a 1990 amendment to the Clean Air Act, the US Environmental Protection Agency (EPA) promulgated the Gasoline Rule on the composition and emissions effects of gasoline, in order to reduce air pollution in the US. From 1 January 1995 (coincidentally the date when the WTO came into being), the Gasoline Rule permitted only gasoline of a specified cleanliness ("reformulated gasoline") to be sold to consumers in the most polluted areas of the country. In the rest of the country, only gasoline no dirtier than that sold in the base year of 1990 ("conventional gasoline") could be sold. The Gasoline Rule applied to all US refiners, blenders and importers of gasoline. It required any domestic refiner which was in operation for at least 6 months in 1990, to establish an individual refinery baseline, which represented the quality of gasoline produced by that refiner in 1990. The Environmental Protection Agency also established a statutory baseline, intended to reflect average US 1990 gasoline quality. The statutory baseline was assigned to those refiners who were not in operation for at least six months in 1990, and to importers and blenders of gasoline. Compliance with the baselines was measured on an average annual basis.

Venezuela and Brazil claimed that the Gasoline Rule was inconsistent, *inter alia*, with GATT Article III ("national treatment"), i.e. treating domestic and imported products equally), and was not covered by GATT Article XX (which deals with general exceptions to the rules, including for certain environmental reasons). The US argued that the Gasoline Rule was consistent with Article III, and, in any event, was justified under the exceptions contained in GATT Article XX, paragraphs (b), (g) and (d).

The Panel found that the Gasoline Rule was inconsistent with Article III, and could not be justified under paragraphs (b), (d) or (g). On appeal of the Panel's findings on Article XX(g), the Appellate Body found that the baseline establishment rules contained in the Gasoline Rule fell within the terms of Article XX(g), but failed to meet the requirements of the "chapeau" (introductory paragraph) of Article XX.

United States — Import Prohibition of Certain Shrimp and Shrimp Products, adopted on 6 November 1998

Seven species of sea turtles have to date been identified. They are distributed around the world in subtropical and tropical areas. They spend their lives at sea, where they migrate between their foraging and nesting grounds. Sea turtles have been adversely affected by human activity, either directly (their meat, shells and eggs have been exploited), or indirectly (incidental capture in fisheries, destruction of their habitats, pollution of the oceans).

The U.S. Endangered Species Act of 1973 (ESA) lists as endangered or threatened the five species of sea turtles occurring in US waters and prohibits their take within the United States, within the US territorial sea and the high seas. Pursuant to the ESA, the United States requires that US shrimp trawlers use "turtle excluder devices" (TEDs) in their nets when fishing in the areas where there is a significant likelihood of encountering sea turtles. Section 609 of Public law 101-102, enacted in 1989 by the United States, provides, inter alia, that shrimp harvested with technology that may adversely affect certain sea turtles may not be imported into the United States, unless the harvesting nation is certified to have a regulatory programme and an incidental take rate comparable to that of the United States, or that the particular fishing environment of the harvesting nation does not pose a threat to sea turtles. In practice, countries having any of the five species of sea turtles within their jurisdiction and harvesting shrimp with mechanical means must impose on their fisherman requirements comparable to those borne by US shrimpers, essentially the use of TEDs at all times, if they want to be certified and export shrimp products to the United States.

India, Malaysia, Pakistan and Thailand complained that the prohibition imposed by the United States on the importation certain shrimp and shrimp products was contrary to Articles I, III and XI of the GATT. The Panel found that the US measure at stake was inconsistent with GATT Article XI (general elimination of quantitative restrictions) and could not be justified under GATT Article XX (general exceptions) because it constituted "unjustifiable discrimination between countries where the same conditions prevailed". Appellate Body found that the measure at stake did qualify for provisional justification under Article XX (g), but failed to meet the requirements of the chapeau of Article XX, and, therefore, was not justified under Article XX of GATT 1994.

3.3 International Treaties on Environment with Trade Provisions

- Convention Relative to the Preservation of Fauna and Flora in their Natural state, 1933

- Convention on Nature Protection and Wildlife Preservation in the Western Hemisphere, 1940

- International Convention for the Protection of Birds, 1950

- International Plant Protection Agreement, 1951

- Plant Protection Agreement for the South East Asia and Pacific Region, 1956

- Convention on Conservation of North Pacific Fur Seals, 1957

- Agreement Concerning the Cooperation in the Quarantine of Plants and Their Protection against Pests and Diseases, 1959

- Phytosanitary Convention for Africa, 1967

- African Convention on the Conservation of Nature and Natural Resources, 1968

- European Convention for the Protection of Animals during International Transport, 1968

- Benelux Convention on the Hunting and Protection of Birds, 1970

- Convention on International Trade in Endangered Species, 1973

- Agreement on the Conservation of Polar Bears, 1973

- Convention for the Conservation and Management of the Vicuna, 1980

- ASEAN Agreement on the Conservation of Nature and Natural Resources, 1985

• International Code of Conduct on the Distribution and Use of Pesticides (UN food and Agricultural Organization), 1985

• Montreal Protocol on Substances That Deplete the Ozone Layer 1987

• Convention for the Prohibition of Fishing with Long Drift Nets in the South Pacific, 1989

• Basel Convention on the Control of Transboundary Movements of Hazardous Wastes and Their Disposal, 1989

• Amended London Guidelines for the Exchange of Information on Chemicals in International Trade, 1989

Chapter 13

GLOBAL ENVIRONMENTAL PROTECTION AND TRADE

In this chapter[1], we will present a theoretical analysis of the participation decisons of countries into an international environmental agreement in the presence of international trade. We will model international environmental agreements (IEAs) as global public good, and the decision to join an IEA will be matched with contributions to the production of this good by each government. Since global protection confers benefits on all countries regardless of their contribution (presence of free-riding, we try to determine (i) how-much each country will contribute, if any; (ii) the factors which affect the level of contributions; (iii) comparison of global environmental protection when countries do not cooperate versus when they cooperate; and finally, (iv) the impact of free trade on these results.

1. Introduction

In an empirical study, Piggott *et al.* (1992) argue that 'the incentives to either participate or not to participate in any international initiatives to reduce carbon emissions reflect much more than free-riding on the benefit side'. A cut in energy use by one country confers benefits to others as well, but the overall welfare assessment of such an action also depends on production substitution across countries, and on terms of trade effects involving energy products, energy-intensive and other products. They show that terms of trade effects associated with reductions in emissions are potentially important, and the numerical values for percentage changes in terms of trade are very high. In our paper,

[1] This chapter is based on my paper titled "Does Trade Always Harm the Global Environment? A Case for Positive Interaction," *Oxford Economic Papers* 50, 272-288; with the permission of Oxford University Press.

S. Alpay (ed.), Trade and The Environment, 183–200.

we will now add theoretical arguments that support Piggott *et al.* findings. We will concentrate on general global environmental protection (not specifically on carbon reduction schemes as in Piggott *et al.* 1992) and its connections with international trade through terms of trade changes.

Global environmental protection should not and can not be studied by itself, in a partial equilibrium framework. Given the connections among countries in the international markets for private goods, it is of interest to determine the impact of international trade on incentives to free ride on global environmental protection. Barrett (1997) develops a stylized model of international trade and environmental cooperation in which the threat to impose trade sanctions emerges as an equilibrium. His model shows that there will not be any trade restriction in equilibrium; however, if the threat to impose trade sanctions is prohibited, for example due to GATT regulations, international cooperation could only sustain a collectively inferior outcome.

Our analysis will highlight one other aspect of the interactions between trade and environmental protection. There are economic links between global environmental protection and international trade, and they need to be examined in a general equilibrium framework. This is so because production of environmental good will be accompanied by pulling up of some resources from the private sector of the economy which engages in the production of goods and services that are internationally traded. Thus, in addition to environmental benefits, there will also be terms of trade and income effects which must be included in the evaluation of environmental policies by governments. As a policy implication, we can state that the assessment of government policies on global environmental protection in a partial equilibrium framework (by ignoring the possible interactions mentioned above) may very well be misleading.

In a study by Hung and Richelle (1997), the connections between international trade and environmental protection has been studied in a framework very similar to ours. They mainly investigate the impact of trade liberalization on the contribution of countries to global protection, the free-riding incentives, the welfare implications and finally the coordinated or cooperative outcome. Their paper considers a special case where there is no substitution possibilities between demand for environmental quality and demand for private goods. We take up a more general case which allows for different substitution possibilities, but this comes at the expense of giving up using a general utility function. The connections between terms of trade changes and environmental protection is more explicitly shown in this paper. Finally contrary to their result on coordinated outcome being resulting in higher environmental protection, in our paper it turns out that the non-cooperative contribution to environmental protection exceeds the one corresponding to the cooperative case.

As mentioned above, Heal (1994) shows that the selection of positive abatement levels by other countries may shift a particular country's best choice of abatement from zero to a positive number if the benefit and cost functions of identical countries are not separable across abatement levels ('reinforcement effects'), or if there are fixed costs associated with abatement programs; thus, free-riding need not always be the best action. In our set-up, we will also be able to see whether Heal's (1994) non-conventional findings will be validated in a general equilibrium model, which incorporates international trade connections among countries. We now introduce our model.

2. Model

We will take up a two-country two-good Ricardian model. Each country consumes two goods, X and Y. There is only one factor of production, labor. Total labor endowment in home country is $\bar{L}$, and the one in foreign country is $\bar{L}^*$. Consumers demand environmental quality along with private consumption goods, and demand for environmental quality is represented by the demand for environmental good, F. Our model will allow for substitution between private goods and environmental quality. The substitution effect is usually ignored in the literature with the exception of a few studies. For example, Smith and Espinosa (1996) point out the importance of this link by stating that 'pollution itself may alter the composition of goods demanded'. In our case, the substitution effect will show itself through the impact of environmental protection on the income, which will be shown explicitly below. Higher demand for environmental quality will attract more resources to environmental sector, and the compensation for these resources will come from taxes paid by consumers; therefore, demand for private goods will decline. Moreover, the environmental agency (or government) chooses the level of environmental protection in a way that maximizes the welfare of the consumers; therefore, the parameters that affect the demand for private goods will also have impact on the level of environmental protection.

2.1 Production side

Production functions are assumed to be linear in labor input. Unit labor requirements are h_x, h_y, and h_f for goods X, Y, and F, respectively, in home country and f_x, f_y, and f_f in foreign country. We assume that full employment of labor prevails in both countries, so $\bar{L} = L_X + L_Y + L_F$ and $\bar{L}^* = L_X^* + L_Y^* + L_F^*$. Throughout our analysis we will assume that these unit labor requirements are such that home country specializes in the production of good X and foreign country in the production of Y. Couple remarks on environmental good F are in order.

In production terms, F can be interpreted as a public good, which represents the cumulative contribution of a country to environmental protection. More explicitly, we formalize the total amount of resources devoted to environmental protection through the production of F. An important nature of the environmental good F is that it not only benefits the country which produces it, but also benefits the other country. Therefore, free-riding will be an important concern in the production of environmental good. In a closed economy, Nash equilibrium production of a good subject to free-riding will be lower than the level corresponding to cooperative solution. Now, the question is whether this result will still be valid in an open economy.

Under the assumed pattern of trade, the incomes of home and foreign country will be $I = P_x \frac{1}{h_x}(\bar{L} - L_F)$ and $I^* = P_y \frac{1}{f_y}(\bar{L}^* - L_F^*)$, respectively. We note that production of F is financed by taxation.

2.2 Demand side

We assume that the representative consumer[2] has the following Cobb-Douglas utility function:

$$U(X, Y, F, F^*) = X^{\alpha} Y^{\gamma} (1 + F + \lambda F^*)^{\beta} \tag{13.1}$$

As usual, α and γ represent, the share of X and Y in total expenditure, respectively, and thus the value of these two goods to the consumer; β plays the same role for environmental quality. Since global environmental protection by any country contributes to the welfare of both countries, utility function includes both F and F^*; however, these contributions are asymmetric as represented by λ; λ measures both the degree of free-riding and the degree of substitution between environmental protection of home country and that of foreign country. Consumer maximizes this utility function subject to the budget constraint, which is $P_x X + P_y Y = I$.

[2]For simplicity, we will assume that there is only one consumer in both home and foreign country, or many consumers with identical utility functions so that we can use a social utility function which is equal to the sum of the individual utility functions, as in Smith and Espinosa (1996).

3. Free-riding and Nash Equilibrium

In this section, we will study the Nash equilibrium of the game where both countries decide whether they should produce environmental good F or not, and if affirmative, how-much they should produce. It is of interest to determine any possible impact of the trade connections between countries on this choice of contribution to environmental protection. Since contribution to global environment requires the transfer of some of the resources from private sector to environmental sector, the interaction between trade and global contribution can be seen by considering the changes in the terms of trade associated with this resource shift. Towards that end, we will determine the Nash equilibrium contributions of the countries when terms of trade changes resulting from the production of F are (i) incorporated and (ii) ignored. We can, then, make an assessment on whether trade connections may reduce free-riding incentives by comparing the level of contributions in each case. Having done these, the natural next step is to show the cooperative solution for this same problem and compare the results with those in the non-cooperative case. With this comparison, we will be able to assess whether the underproduction of the public goods in the closed economy will extend to the open economy case (when there are trade connections). After this overview, we can now see the details.

One may separate the overall process into two stages. In the first stage, the representative consumer maximizes his/her utility over the private goods for a given level of domestic and foreign environmental protection, F and F^*, respectively. In the next stage, domestic environmental agency chooses the amount of F such that the utility of the representative consumer as computed in the first stage is maximized. The consumer's problem will be as follows:

$$\max_{X,Y} \; U(X, Y, F, F^*) = X^{\alpha} \, Y^{\gamma} \, (1 + F + \lambda F^*)^{\beta}$$

subject to the budget constraint, $P_x X + P_y Y = I$, where $I = P_x \frac{\bar{L} - L_F}{h_x}$. Allocation of some of the resources to environmental protection (L_F) will affect the private income as reflected in the budget equation.

The resulting demand functions are:

$$X = \frac{\alpha}{(\alpha + \gamma)} \frac{I}{P_x} = \frac{\alpha}{(\alpha + \gamma)} \frac{(\bar{L} - L_F)}{h_x} \tag{13.2}$$

$$Y = \frac{\gamma}{(\alpha + \gamma)} \frac{I}{P_y} = \frac{\gamma}{(\alpha + \gamma)} \frac{P_x}{P_y} \frac{(\bar{L} - L_F)}{h_x} \tag{13.3}$$

Since foreign country specializes in the production of Y, the demand functions for their representative consumer will be the symmetric of the ones corresponding to the home country above:

$$X^* = \frac{\alpha}{(\alpha+\gamma)} \frac{P_y}{P_x} \frac{(\bar{L}^* - L_F^*)}{f_y} \tag{13.4}$$

$$Y^* = \frac{\gamma}{(\alpha+\gamma)} \frac{(\bar{L}^* - L_F^*)}{f_y} \tag{13.5}$$

Trade balance requires that export receipts will be same as import payments; since home country imports good Y and exports good X, the following relation must hold:[3]

$$P_y Y = P_x X^* \tag{13.6}$$

Replacing demand functions, we obtain the terms of trade as:

$$\frac{P_x}{P_y} = \frac{\alpha}{\gamma} \frac{h_x}{f_y} \frac{(\bar{L}^* - L_F^*)}{(\bar{L} - L_F)} \tag{13.7}$$

Given the optimal choices of X and Y, and the trade balance equation, the representative consumer's maximum utility can be expressed as a function of L_F and L_F^* as:

$$U(X, Y, F, F^*) = \frac{\alpha^{\alpha+\gamma}}{(\alpha+\gamma)^{\alpha+\gamma}} \frac{1}{h_x^\alpha f_y^\gamma} (\bar{L}^* - L_F^*)^\gamma (\bar{L} - L_F)^\alpha (1 + \frac{L_F}{h_f} + \lambda \frac{L_F^*}{f_f})^\beta \tag{13.8}$$

Now, we are at the second stage. The environmental agency chooses L_F such that the utility of the consumer expressed in (13.8) reaches a maximum. The problem of the agency is:

$$\max_{L_F} \frac{\alpha^{\alpha+\gamma}}{(\alpha+\gamma)^{\alpha+\gamma}} \frac{1}{h_x^\alpha f_y^\gamma} (\bar{L}^* - L_F^*)^\gamma (\bar{L} - L_F)^\alpha (1 + \frac{L_F}{h_f} + \lambda \frac{L_F^*}{f_f})^\beta$$

From the solution of first order conditions, we obtain:

$$L_F = \frac{\beta}{\alpha+\beta} \bar{L} - \frac{\alpha}{\alpha+\beta} h_f - \frac{\alpha \lambda h_f}{(\alpha+\beta) f_f} L_F^* \tag{13.9}$$

[3]Note that trade balance together with budget constraints results in the equality of demand and supply worldwide, i.e., $X + X^* = X_S$, and $Y + Y^* = Y_S$, where X_S and Y_S represent supply of goods.

Equation (13.9) is the reaction function of home country with respect to L_F^* of foreign country. A similar analysis for foreign country results in the following reaction function for foreign country with respect to L_F of home country:

$$L_F^* = \frac{\beta}{\gamma+\beta}\bar{L}^* - \frac{\gamma}{\gamma+\beta}f_f - \frac{\gamma\lambda f_f}{(\gamma+\beta)h_f}L_F \tag{13.10}$$

To determine the Nash equilibrium of this game, we first calculate the welfare levels associated with each of the four possible outcomes, by using the optimal contributions above in (13.9) and (13.10) when they choose to contribute:

Case I. Both countries contribute:

$$U(X,Y,F,F^*) = \frac{\alpha^{\alpha+\gamma}}{(\alpha+\gamma)^{\alpha+\gamma}}\frac{1}{h_x^\alpha\, f_y^\gamma}(\bar{L}^*-L_F^*)^\gamma(\bar{L}-L_F)^\alpha(1+\frac{L_F}{h_f}+\lambda\frac{L_F^*}{f_f})^\beta \tag{13.11}$$

$$U^*(X,Y,F,F^*) = \frac{\gamma^{\alpha+\gamma}}{(\alpha+\gamma)^{\alpha+\gamma}}\frac{1}{h_x^\alpha\, f_y^\gamma}(\bar{L}^*-L_F^*)^\gamma(\bar{L}-L_F)^\alpha(1+\frac{L_F^*}{f_f}+\lambda\frac{L_F}{h_f})^\beta \tag{13.12}$$

Case II. None of the countries contributes:

$$U(X,Y,F,F^*) = \frac{\alpha^{\alpha+\gamma}}{(\alpha+\gamma)^{\alpha+\gamma}}\frac{1}{h_x^\alpha\, f_y^\gamma}\bar{L}^{*\gamma}\,\bar{L}^\alpha \tag{13.13}$$

$$U^*(X,Y,F,F^*) = \frac{\gamma^{\alpha+\gamma}}{(\alpha+\gamma)^{\alpha+\gamma}}\frac{1}{h_x^\alpha\, f_y^\gamma}\bar{L}^{*\gamma}\,\bar{L}^\alpha \tag{13.14}$$

Case III. Home country contributes, foreign country does not:

$$U(X,Y,F,F^*) = \frac{\alpha^{\alpha+\gamma}}{(\alpha+\gamma)^{\alpha+\gamma}}\frac{1}{h_x^\alpha\, f_y^\gamma}\bar{L}^{*\gamma}\,(\bar{L}-L_F)^\alpha\,(1+\frac{L_F}{h_f})^\beta \tag{13.15}$$

$$U^*(X,Y,F,F^*) = \frac{\gamma^{\alpha+\gamma}}{(\alpha+\gamma)^{\alpha+\gamma}}\frac{1}{h_x^\alpha\, f_y^\gamma}\bar{L}^{*\gamma}\,(\bar{L}-L_F)^\alpha\,(1+\lambda\frac{L_F}{h_f})^\beta \tag{13.16}$$

Case IV. Foreign country contributes, home country does not:

$$U(X,Y,F,F^*) = \frac{\alpha^{\alpha+\gamma}}{(\alpha+\gamma)^{\alpha+\gamma}}\frac{1}{h_x^\alpha\, f_y^\gamma}(\bar{L}^*-L_F^*)^\gamma\,\bar{L}^\alpha\,(1+\lambda\frac{L_F^*}{f_f})^\beta \tag{13.17}$$

$$U^*(X,Y,F,F^*) = \frac{\gamma^{\alpha+\gamma}}{(\alpha+\gamma)^{\alpha+\gamma}}\frac{1}{h_x^\alpha\, f_y^\gamma}(\bar{L}^*-L_F^*)^\gamma\,\bar{L}^\alpha\,(1+\frac{L_F^*}{f_f})^\beta \tag{13.18}$$

Given these pay-offs with many different parameters, we need to make some simplifying assumptions to be able to track down the Nash equilibrium of this game. One interesting specification is the case of symmetric countries, a very common assumption in this literature.

Definition The countries are symmetric if $\bar{L} = \bar{L}^*$, $h_f = f_f$, and $\alpha = \beta = \gamma$.

Then, our first result is:

Theorem 1 If the countries are symmetric, and if the following two conditions (i) $\left(\frac{1+\lambda}{2+\lambda} - \frac{\lambda}{2h_f}\right)\bar{L} > \frac{2-\lambda}{2} - \frac{1}{2+\lambda}h_f$, (ii) $h_f < \frac{1+\lambda}{2+\lambda}$ are satisfied, then the Nash equilibrium of the game described above does not support free-riding i.e., both countries choose to contribute to global protection.

Proof of Theorem 1 is given in appendix. Note that the conditions of this theorem are not necessary conditions; there may be many other cases, in which these conditions will not be satisfied, but the result of Theorem 1 will still hold.

A common view in the literature of global environmental issues is that co-operation for the protection of global environment is very difficult to obtain, and even if it is realized, every country will earn higher benefits by free-riding on the contribution of other countries (see Demsetz, 1967, for example). This view can not always be true. Our last theorem shows that due to interactions in international goods markets, free-riding is not always the best policy for the countries. Even in a non-cooperative setting, contribution to global protection brings a higher return for both countries under some conditions. Hence, the reluctance to undertake pro-environmental policies due to free-riding problem is not grounded solidly in an open economy. Let us try to understand the reasons behind this unconventional result.

3.1 Terms of Trade Effects

When both countries start allocating some of their resources in the production of environmental good, the resources available for production in the private sector will decline. As a result, such environmental policies will have terms of trade effects, the magnitude of which should be a function of the resources transferred to environmental sector. This relationship is shown in equation (13.7) above.

In the public good literature, it is known that the agents would like to free ride and the production of public good will be suboptimal, i.e., public good will be under-produced. This is a conventional result in a closed economy. When the analog of this public good problem in an open economy is desired to be developed, one need to consider the trade related side effects. The obvious side effect is the changes in the terms of trade generated by the production of public good. Let us state this more formally:

Lemma In an open economy, along with the positive environmental externality, the production of environmental good also generates negative externality on the free-riding country and positive externality on the active country through terms of trade changes.

Proof First we note that the welfare of home country is positively related to the terms of trade, $\frac{P_x}{P_y}$, which can be seen by considering equations (13.1) and (13.3) together. A similar consideration will show that the utility function of the representative consumer in foreign country is negatively related to the same ratio. The relationship between environmental good production (which is a function of L_F), and the terms of trade is given in equation (13.7). The derivative of the terms of trade with respect to L_F will be (from 13.7):

$$\frac{d\frac{P_x}{P_y}}{dL_F} = \frac{\alpha}{\gamma}\frac{h_x}{f_y}\frac{\bar{L}^* - L_F^*}{(\bar{L} - L_F)^2} \tag{13.19}$$

As this derivative is positive, higher environmental protection increases terms of trade to the favour of home country[4] and thus, it rises the welfare of home country and causes a decline in the welfare of foreign country. The lemma is thus proved. Also, note that $\left(\frac{d(P_x/P_y)}{dL_F^*}\right)$ is negative and this shows that the indirect welfare effects of the foreign environmental good production (due to terms of trade changes) are positive for foreign country and negative for home country.

Since terms of trade changes associated with environmental protection generate welfare effects, they also affect the total contribution to global environment by each country. This can be shown more explicitly by considering the case in which the terms of trade is kept constant before and after the environmental policy. As the base case, consider that none of the countries produce environmental good, and thus, the terms of trade will be

$$\frac{P_x}{P_y} = \frac{\alpha}{\gamma}\frac{h_x}{f_y}\frac{\bar{L}^*}{\bar{L}} \tag{13.20}$$

[4]Recall that home country exports X and imports Y.

Now, assume that both countries contribute to environmental protection, but we keep the terms of trade constant at its level in (13.20). Incorporating this, optimal choices of L_F and L_F^* will be obtained from the solution of:

$$\max_{L_F} \frac{\alpha^{\alpha+\gamma}}{(\alpha+\gamma)^{\alpha+\gamma}} \frac{1}{h_x^{\alpha} f_y^{\gamma}} \left(\frac{\bar{L}^*}{\bar{L}}\right)^{\gamma} (\bar{L} - L_F)^{\alpha+\gamma} (1 + \frac{L_F}{h_f} + \lambda \frac{L_F^*}{f_f})^{\beta}$$

and the resulting reaction functions are:

$$L_F^{N_p} = \frac{\beta}{\alpha+\gamma+\beta}\bar{L} - \frac{\alpha+\gamma}{\alpha+\gamma+\beta}h_f - \frac{(\alpha+\gamma)\lambda h_f}{(\alpha+\gamma+\beta)f_f}L_F^{*N_p} \tag{13.21}$$

where N_p in $L_F^{N_p}$ refers to non-cooperative contribution when the terms of trade changes are ignored. Similarly,

$$L_F^{*N_p} = \frac{\beta}{\alpha+\gamma+\beta}\bar{L}^* - \frac{\alpha+\gamma}{\alpha+\gamma+\beta}f_f - \frac{(\alpha+\gamma)\lambda f_f}{(\alpha+\gamma+\beta)h_f}L_F^{N_p} \tag{13.22}$$

Now, let us try to compare the total contributions here with the ones obtained above for the case when terms of trade changes were taken into consideration. This is our next result:

Theorem 2 Under the assumptions of Theorem 1, the overall Nash equilibrium contributions of the countries will be under-estimated if the terms of trade changes associated with environmental protection are ignored.

Proof By using the symmetry assumption of Theorem 1, we can simplify the reaction functions in (13.21) and (13.22) to obtain (note that, in this case β may be different from α and γ)

$$L_{Total}^{N_p} = \frac{\beta}{2\alpha(1+\lambda)+\beta}(\bar{L}+\bar{L}^*) - \frac{4\alpha}{2\alpha(1+\lambda)+\beta}h_f \tag{13.23}$$

and from equations (13.9) and (13.10)

$$L_{Total}^{N} = \frac{\beta}{\alpha(1+\lambda)+\beta}(\bar{L}+\bar{L}^*) - \frac{2\alpha}{\alpha(1+\lambda)+\beta}h_f \tag{13.24}$$

where N in L_F^N refers to non-cooperative contribution when terms of trade changes are incorporated as in the previous section.

Since $\frac{\beta}{2\alpha(1+\lambda)+\beta} < \frac{\beta}{\alpha(1+\lambda)+\beta}$ and $\frac{4\alpha}{2\alpha(1+\lambda)+\beta} > \frac{2\alpha}{\alpha(1+\lambda)+\beta}$, it follows from equations (13.23) and (13.24) that $L_{Total}^{N} > L_{Total}^{N_p}$.

The intuition behind this result is as follows: terms of trade changes affect the marginal product of resources being used in the environmental sector positively;

therefore more labor moves to the environmental sector, and so, environmental protection by each country becomes higher when the changes in the prices of private goods are incorporated. This shows that the indirect trade effects of environmental protection are as important as the direct effects. As another important interpretation of Theorem 2, we note that trade connections among countries reduce the free-riding incentives on global environmental protection of others. This enhances the value of trade liberalization, this time, from an environmental perspective.

It is also important to note that the reaction functions described above incorporate a number of opposing effects: (i) transboundary nature of the problem which tends to cause free-riding and thus lowers environmental protection; (ii) trade connections generate terms of trade benefits which tend to increase contribution to environmental protection; (iii) the substitution possibilities between environmental good and private goods tend to lower environmental protection, as lower contribution means higher private income, and so, higher private good consumption. In our case, the last effect is dominant to others and so, reaction functions are always negatively sloped. This also shows the importance of the incorporation of substitution possibilities between private goods and environmental good into the model; most of the other studies in this literature ignore this.

Now, having studied the non-cooperative solutions, the natural next step is to obtain the cooperative solution. How does overall cooperative contribution compare to the non-cooperative one?

3.2 Cooperative Case

Let us assume that the contributions of each country to global environmental protection are determined by a social planner. The role of the social planner is to internalize the production externalities associated with the production of F and F^* given the optimal behaviors of the representative consumer in each country. This will be realized through the following maximization problem:

$$\max_{F,F^*} \; U(X_o, Y_o, F, F^*) + U^*(X_o^*, Y_o^*, F, F^*)$$

where X_o, Y_o, X_o^*, and Y_o^* are optimal choices of the private goods. When the sum of the welfare levels of the two countries are maximized by the social planner, one may reasonably argue that the solution to the social planner's optimization problem above may result in a lower welfare for one of the countries as compared to its welfare level in the non-cooperative case. Such possibilities must be taken into consideration. The results of this section are independent of such cases, and it can be shown that these important results will hold in cases where both countries obtain higher welfare in the cooperative case as compared to the non-cooperative case.

Replacing X_o, Y_o, X_o^*, and Y_o^*, by using our results above, the open form of the social planner's problem is

$$\max_{L_F, L_F^*} \frac{\alpha^{\alpha+\gamma}}{(\alpha+\gamma)^{\alpha+\gamma}} \frac{1}{h_x^\alpha f_y^\gamma} (\bar{L}^* - L_F^*)^\gamma (\bar{L} - L_F)^\alpha (1 + \frac{L_F}{h_f} + \lambda \frac{L_F^*}{f_f})^\beta + \frac{\gamma^{\alpha+\gamma}}{(\alpha+\gamma)^{\alpha+\gamma}} \frac{1}{h_x^\alpha f_y^\gamma} (\bar{L}^* - L_F^*)^\gamma (\bar{L} - L_F)^\alpha (1 + \frac{L_F^*}{f_f} + \lambda \frac{L_F}{h_f})^\beta \quad (13.25)$$

To be able to solve the first order conditions of the above maximization problem, we impose that $\beta = 1$. Then,

$$L_F^C = \frac{1}{\alpha+1}\bar{L} - \frac{2\alpha}{(1+\lambda)(\alpha+1)}h_f - \frac{\alpha h_f}{(\alpha+1)f_f}L_F^{*C} \quad (13.26)$$

where C in L_F^C refers to cooperative solution; also, note that,

$$L_F^{*C} = \frac{1}{\alpha+1}\bar{L}^* - \frac{2\alpha}{(1+\lambda)(\alpha+1)}f_f - \frac{\alpha f_f}{(\alpha+1)h_f}L_F^C \quad (13.27)$$

The comparison of the overall contribution to global protection in the cooperative case and the one in the non-cooperative case generates our next result:

Theorem 3 Under the assumptions of Theorem 1, if $\beta = 1$, the overall contribution of countries to global environmental protection in the cooperative case will be smaller than their contributions in the non-cooperative case when terms of trade changes are taken into consideration.

Proof Using equations (13.26) and (13.27) it can easily be shown that

$$L_{Total}^C = \frac{1}{2\alpha+1}(\bar{L} + \bar{L}^*) - \frac{4\alpha}{2\alpha(\lambda+1)+\lambda+1}h_f \quad (13.28)$$

The overall contribution in the non-cooperative case (including terms of trade changes) is given by equation (13.24) above. Substitute 1 for β in (13.24). Note that we do not need the assumption $\alpha = \gamma = \beta$ here. Since $\frac{1}{\alpha(1+\lambda)+1} > \frac{1}{2\alpha+1}$ and $\frac{2\alpha}{\alpha(1+\lambda)+1} < \frac{4\alpha}{2\alpha(\lambda+1)+\lambda+1}$, then $L_{Total}^N > L_{Total}^C$, which is our last theorem.

This result is opposite to the common belief that public goods will be undersupplied in the non-cooperative case as compared to the cooperative case. The main reason for this result is the indirect effects of environmental good production on the welfare levels through changes in the internationally traded goods market. When countries contribute to global environmental protection, they pull away some of their resources from private sector which engages in trade with other countries, and this generates terms of trade effects. If the pattern of specialization remains the same after the pro-environmental policy, the terms of trade of the country producing environmental good improves and this increases the returns to employing labor in the production of environmental good. Thus, more labor moves into environmental sector as compared to the case when there is no terms of trade effect (or when the terms of trade changes are ignored). In the cooperative case, since the sum of the welfare levels of both countries are maximized, terms of trade effects cancel each other, as an improvement in one country's terms of trade is equivalent to a worsening in the other.

To reinforce this result, we should compare the cooperative result with the non-cooperative result that ignores terms of trade changes. As we show now, the traditional result that the non-cooperative contribution of countries is smaller than their cooperative contributions, holds when terms of trade changes are ignored.

Corollary 1 Under the assumptions of Theorem 3, the production of environmental good will be larger in the cooperative case than in the non-cooperative case if the terms of trade changes associated with the production of environmental good are ignored.

Proof Firstly, we replace β in (13.23) by 1 and get the total contribution to environmental protection in the non-cooperative case, in which terms of trade changes are ignored, as:

$$L_{Total}^{N_p} = \frac{1}{2\alpha(1+\lambda)+1}(\bar{L}+\bar{L}^*) - \frac{4\alpha}{2\alpha(1+\lambda)+1}h_f \tag{13.29}$$

Comparison of L_{Total}^{C}, total cooperative contribution to global protection, in (13.28) with $L_{Total}^{N_p}$ in (13.29) clearly shows that $L_{Total}^{C} > L_{Total}^{N_p}$.

Combining this final result with our previous results, we can state that $L_{Total}^{N} > L_{Total}^{C} > L_{Total}^{N_p}$, that is to say, global environmental protection will be higher in the non-cooperative case than in the cooperative case when terms of trade changes are taken into consideration, which is contrary to the conventional results. We get the conventional result, i.e. lower environmental protection in the non-cooperative case, when terms of trade changes are ignored. Thus, we conclude that partial equilibrium analysis of global environmental protection which ignores trade connections among countries, may be very misleading.

In an open economy, the public good problem that the agents tend to free-ride on the contribution of others, does not hold when the environmental externality can not overcome the terms of trade externality associated with the public good production. In a larger picture, it seems possible that, even in a domestic economy, if the contributions of the agents to public good production generate positive side effects for them, then they may also choose to contribute voluntarily; however, since the individuals are most of the time negligible in the total, their decisions will not generally bring about spillovers that may change their incentives to contribute. In our open economy case, however, the countries are not negligible, and the terms of trade benefits arising from environmental protection follow, and through this channel, countries' contributions exceed the optimal contributions associated with the cooperative case, which is a non-conventional result.

4. Conclusion

This chapter mainly takes up the question whether the suboptimality of the production of a public good in a closed economy extends to open economy case, more explicitly, whether global environmental protection is also subject to free-riding in a model where trade connections among countries are taken into consideration. By studying the case of contributions to global environmental protection in a 2x2 Ricardian trade model, we demonstrate that there are links between international trade and environmental control, heretofore unappreciated, which might substantially alter the efficacy of various governmental policies to control pollution. One concern about national environmental policies is that, whereas the benefits of certain types of abatement might be international or even worldwide, the costs will be borne strictly by the consumers and firms of the country which institutes the policy. As a result, for those types of pollution which are global (such as greenhouse gasses) there will be 'too little' pollution abatement. This pessimistic conclusion might be unwarranted. If the abatement is carried out by diverting resources from export sector, terms of trade of the country engaged in abatement will improve, with the result that the country is not so reluctant to clean up the environment after all.

It is also demonstrated that there are cases in which contribution to global environmental protection by both countries is the non-cooperative Nash equilibrium of this environmental game. Again contrary to the common belief, the corresponding environmental good production in the non-cooperative case will exceed the one associated with the cooperative solution, under reasonable assumptions. Our results also imply that partial equilibrium studies on free-riding incentives on global protection which ignore the trade connections among countries may very well be misleading. Incorporation of trade connections turns out to be very important. As reflected in our results, the impact of pro-environmental policy on the terms of trade can outweigh the environmental gains, and thus free-riding may not be a welfare-improving choice. As a policy implication, this requires that the governments consider all aspects of their environmental policies, at least the ones that pertain to international trade.

Chilchilnisky (1994b) states that 'the international market transmits and enlarges the externalities of the global commons. No policy that ignores this connection can work'. We agree with her results; however, we show that international market can also transmit and enlarge global environmental protection across countries by reducing free-riding incentives. Thus, trade is not always a threat to global environment.

5. Appendix

Proof of Theorem 1

We can prove this theorem by showing that contributing to global environmental protection is dominant strategy for both countries. Let us take home country first. Assume that foreign country chooses to contribute, then is it the case that home country will have higher welfare if it also contributes? This requires that the welfare of home country is greater in case I (both countries contribute) than in case IV (foreign country contributes, and home country does not), that is to say,

$$\frac{\alpha^{\alpha+\gamma}}{(\alpha+\gamma)^{\alpha+\gamma}}\frac{1}{h_x^{\alpha}\, f_y^{\gamma}}\,(\bar{L}^* - L_F^{*I})^{\gamma}\,(\bar{L} - L_F^{I})^{\alpha}\,(1 + \frac{L_F^{I}}{h_f} + \lambda\frac{L_F^{*I}}{f_f})^{\beta} >$$
$$\frac{\alpha^{\alpha+\gamma}}{(\alpha+\gamma)^{\alpha+\gamma}}\frac{1}{h_x^{\alpha}\, f_y^{\gamma}}\,(\bar{L}^* - L_F^{*IV})^{\gamma}\,(\bar{L})^{\alpha}\,(1 + \lambda\frac{L_F^{*IV}}{f_f})^{\beta} \quad (13.30)$$

where the superscripts I and IV refer to case I and case IV, respectively. Making possible cancellations, and noting that $\alpha = \gamma = \beta$, we are left with:

$$(\bar{L}^* - L_F^{*I})(\bar{L} - L_F^{I})\,(1 + \frac{L_F^{I}}{h_f} + \lambda\frac{L_F^{*I}}{f_f}) > (\bar{L}^* - L_F^{*IV})\,(\bar{L})\,(1 + \lambda\frac{L_F^{*IV}}{f_f}) \quad (13.31)$$

From equation (13.9), which gives the optimal value of L_F as a function of L_F^*, we can show that

$$L_F = \frac{1}{2+\lambda}\bar{L} - \frac{1}{2+\lambda}h_f$$

under symmetry assumptions. Thus, $L_F < \bar{L}$, and $L_F > 0$ if $\bar{L} > h_f$; this condition should always be satisfied as $\bar{L}$ represents total labor force and h_f is the unit labor requirement in the production of environmental good. So, contribution to global protection (L_F) is feasible and positive. The following useful derivations can easily be done now (note that $L_F^{I} = L_F^{*I}$ due to symmetry):

$$(1 + \frac{L_F^{I}}{h_f} + \lambda\frac{L_F^{*I}}{f_f}) = \frac{1+\lambda}{(2+\lambda)h_f}\bar{L} + \frac{1}{2+\lambda}$$

$$\bar{L} - L_F^{I} = \frac{1+\lambda}{2+\lambda}\bar{L} + \frac{1}{2+\lambda}h_f$$

and from equation (13.10), which gives the value of L_F^* as a function of L_F, we get

$$1 + \lambda\frac{L_F^{*IV}}{f_f} = \frac{\lambda}{2h_f}\bar{L} + \frac{2-\lambda}{2}$$

by noting that $\beta = \gamma$, $\bar{L}^* = \bar{L}$ and $h_f = f_f$ (due to symmetry), and that $L_F^{IV} = 0$ since case IV refers to no contribution by home country. Thus, (13.31) simplifies to

$$(\bar{L}^* - L_F^{*I})\,(\frac{1+\lambda}{2+\lambda}\bar{L} + \frac{1}{2+\lambda}h_f)\,(\frac{1+\lambda}{(2+\lambda)h_f}\bar{L} + \frac{1}{2+\lambda}) > (\bar{L}^* - L_F^{*IV})\,\bar{L}\,(\frac{\lambda}{2h_f}\bar{L} + \frac{2-\lambda}{2}) \quad (13.32)$$

The first term on the left-hand side of equation (13.32) is larger than the first term on the right-hand side since $L_F^{*I} < L_F^{*IV}$ (see equation (13.10) with $L_F^{IV} = 0$ and $L_F^I > 0$). The third term on the LHS is larger than the second term on the RHS if $h_f < \frac{1+\lambda}{2+\lambda}$, which is assumed to hold in Theorem 1. Finally, Theorem 1 assumes that the second term on the LHS is larger than the last term on the RHS, i.e., $\frac{1+\lambda}{2+\lambda}\bar{L} + \frac{1}{2+\lambda}h_f > \frac{\lambda}{2h_f}\bar{L} + \frac{2-\lambda}{2}$. Thus, LHS will be larger than RHS in equation (13.30) and so, home country will choose to contribute.

Now, we need to show that contribution will again be dominant for home country when foreign country does not contribute. This requires that the utility of home country is greater in case III (home country contributes, and foreign country does not) than in case II (none contributes), i.e.,

$$\frac{\alpha^{\alpha+\gamma}}{(\alpha+\gamma)^{\alpha+\gamma}}\,\frac{1}{h_x^\alpha\, f_y^\gamma}\,\bar{L}^{*\gamma}\,(\bar{L} - L_F)^\alpha\,(1 + \frac{L_F}{h_f})^\beta > \frac{\alpha^{\alpha+\gamma}}{(\alpha+\gamma)^{\alpha+\gamma}}\,\frac{1}{h_x^\alpha\, f_y^\gamma}\,\bar{L}^{*\gamma}\,\bar{L}^\alpha \quad (13.33)$$

which simplifies to,

$$(\bar{L} - L_F)(1 + \frac{L_F}{h_f}) > \bar{L} \quad (13.34)$$

Left-hand side of (13.34) represents no contribution by foreign country, i.e., $L_F^* = 0$. Putting this in the optimal value of L_F in (13.9) and using $\beta = \alpha$, we get

$$\bar{L} - L_F = \frac{1}{2}\bar{L} + \frac{1}{2}h_f$$

$$1 + \frac{L_F}{h_f} = \frac{1}{2}\frac{\bar{L}}{h_f} + \frac{1}{2}$$

Substituting these in equation (13.34),

$$\frac{1}{4h_f}(\bar{L} + h_f)^2 > \bar{L}$$

which simplifies to

$$(\bar{L} - h_f)^2 > 0$$

which is always true. Again contribution is dominant strategy for home country.

Since countries are symmetric, contribution to global protection will be dominant strategy for foreign country too. Thus, in the Nash equilibrium both countries will choose to contribute to global environmental protection.

A brief note on the derivation of reaction functions in the cooperative case:

The maximization problem given in equation (13.25) results in the following first order conditions with respect to L_F of home country:

$$A\left\{-\alpha(\bar{L}-L_F)^{\alpha-1}(1+\frac{L_F}{h_f}+\lambda\frac{L_F^*}{f_f})^{\beta}+\frac{\beta}{h_f}(\bar{L}-L_F)^{\alpha}(1+\frac{L_F}{h_f}+\lambda\frac{L_F^*}{f_f})^{\beta-1}\right\}$$

$$+B\left\{-\alpha(\bar{L}-L_F)^{\alpha-1}(1+\lambda\frac{L_F}{h_f}+\frac{L_F^*}{f_f})^{\beta}+\frac{\lambda\beta}{h_f}(\bar{L}-L_F)^{\alpha}(1+\lambda\frac{L_F}{h_f}+\frac{L_F^*}{f_f})^{\beta-1}\right\}=0$$

where $A=\frac{\alpha^{\alpha+\gamma}}{(\alpha+\gamma)^{\alpha+\gamma}}\frac{1}{h_x^{\alpha}f_y^{\gamma}}(\bar{L}^*-L_F^*)^{\gamma}$ and $B=\frac{\gamma^{\alpha+\gamma}}{(\alpha+\gamma)^{\alpha+\gamma}}\frac{1}{h_x^{\alpha}f_y^{\gamma}}(\bar{L}^*-L_F^*)^{\gamma}$.

A similar expression will result when derivative of (13.25) is taken with respect to L_F^*. As seen clearly, it is difficult to solve for L_F in the above expression; however, if we let $\beta=1$ as in theorem 3, we will be able to solve for reaction functions under the usual symmetry assumptions. The resulting reaction functions are given in equations (13.26) and (13.27) in section 3.

References

Abler, D.G., Rodriguez, A.G., Shortle, J.S., 1999, Trade Liberalization and the Environment in Costa Rica, *Environment and Development Economics* 4(3), 357-73.

Abrego, E.L., Perroni, C., Whalley, J., and Randall, M.W., 1999, Trade and Environment: Bargaining Outcomes from Linked Negotiations, *CSGR Working Paper*, No. 27, University of Warwick.

Agras, J. and Chapman, D., 1999, A Dynamic Approach to the Environmental Kuznets Curve Hypothesis, *Ecological Economics* 28, 267-277.

Albrecht, J.A., 1998, Environmental Costs and Competitiveness: A Product-Specific Test of the Porter Hypothesis, manuscript, www.ssrn.com .

Alpay, S., 1999, "How Can Trade Liberalization Be Conducive to a Better Environment? A Survey of the Literature", mimeo, Bilkent University.

Alpay, S., 2000, Does Trade Always Harm the Global Environment? A Case for Positive Interaction, *Oxford Economic Papers* 50, 272-288.

Alpay, S., A. Caliskan, and S. Mahmud, 2001, "The Interactions Among Environmental Performance, Income and Openness: A Direct Empirical Analysis of Environmental Kuznets Curve," mimeo, Bilkent University.

Alpay, S., and S. Karaman, 2002, "Trade, R&D and Environmental Quality," mimeo, Bilkent University.

Alpay, S., and M. Caner, 2002, "Environmental Kuznets Curve (EKC) Hypothesis Revisited," mimeo, Bilkent University.

Anderson, K., 1992a, The Standard Welfare Economics of Policies Affecting Trade and the Environment, *in* The Greening of World Trade Issues, (K. Anderson, and R. Blackhurst eds.), Ann Arbor: University of Michigan Press.

Anderson, K., 1992b, Effects on the Environment and Welfare of Liberalizing World Trade: The Cases of Coal and Food, *in* The Greening of World Trade Issues, (K. Anderson, and R. Blackhurst eds.), Ann Arbor: University of Michigan Press.

Anderson, K., and R. Blackhurst, 1992, The Greening of World Trade Issues, Ann Arbor: University of Michigan Press.

Antweiler, W., B.R. Copeland, and M.S.Taylor, 2001, Is Free Trade Good for the Environment, *American Economic Review*, 91, 877-908.

Arnold, M.B. and E.J. Long, 1994, Trade and the Environment: Challenges and opportunities, *in* Environmental Strategies Handbook, (R. V. Kolluru ed.), New York: McGraw-Hill.

Asako, K, 1979, "Environmental Pollution in an Open Economy," *Economic Record*, 55, 359-67.

Bac, M., 1996, Incomplete Information and Incentives to Free Ride on International Environmental Resources, *Journal of Environmental Economics and Management*, 30(3), 301-15.

Barrett, S., 1990, The Problem of Global Environmental Protection, *Oxford Review of Economic Policy*, 6(1), 68-79.

Barrett, S., 1994a, Self-Enforcing International Environmental Agreements, *Oxford Economics Papers*, 46, 878-94.

Barrett, S., 1994b, Strategic Environmental Policy and International Trade, *Journal of Public Economics*, 54, 325-338.

Barrett, S., 1997, The Strategy of Trade Sanctions in International Environmental Agreements, *Resource and Energy Economics* 19, 345-61.

Barrett, S., 1999, The Credibility of Trade Sanctions in International Environmental Agreements, *in* Trade, Global Policy and the Environment, (P.G. Fredriksson ed.), The World Bank, Washington, DC.

Bartik, T.J., 1988, The Effects of Environmental Regulation on Business Location in the United States, *Growth and Change*, 19(3), 22-44.

Bartik, T.J., 1989, Small Business Start-Ups in the United States: Estimates of the Effects of Characteristics of States, *Southern Economic Journal*, 55(4), 1004-18.

Batra, R., Beladi, H., Frasca, R., 1998, Environmental Pollution and World Trade, *Ecological Economics*, 27, 171-182.

Baumol, W.J., 1971, *Environmental Protection, International Spillovers, and Trade*, Almquit and Wixell, Stockholm.

Baumol, W.J., and W.E. Oates, 1988, The Theory of Environmental Policy, Cambridge: Cambridge University Press.

Beghin, J., Roland-Holst, D., van-der-Mensbrugghe, D., 1994, A Survey of the Trade and Environment Nexus: Global Dimensions, *OECD-Economics Studies*, Winter, 167-92.

Beghin, J., Dessus, S., Roland-Holst, D., van-der-Mensbrugghe, D., 1997, The Trade and Environment Nexus in Mexican Agriculture, *Agricultural Economics*, 17(2-3), 115-131.

Beghin, J., Bowland, B., Dessus, S., Roland-Holst, D., van-der-Mensbrugghe, D., 1999, Trade Integration, Environmental Degradation, and Public Health in Chile: Assessing the Linkages, In: Fredriksson, P.G., (Ed.), *Trade, Global Policy and the Environment*, The World Bank, Washington, DC.

Bhagwati, J., 1993, The Case For Free Trade, *Scientific American* November, 18-23.

Birdsall, N. and Wheeler, D., 1992, Trade Policy and Industrial Pollution in Latin America: Where are the Pollution Havens? *in* International Trade and the Environment, (P. Low ed.), World Bank, Washington, DC.

Blackhurst, R., 1977, International Trade and Domestic Environmental Policies in a Growing world Economy, *in* International Relations in a Changing World, (R. Blackhurst, et al. eds.), Geneva: Sythoff-Leiden.

Blackhurst, R., and A. Subramanian, 1992, Promoting Multilateral Cooperation on the Environment, *in* The Greening of World Trade Issues, (K. Anderson, and R. Blackhurst eds.), Ann Arbor: University of Michigan Press.

Bohm, P., 1972, Estimating Demand for Public Goods: An Experiment, *European Economic Review*, 3, 111-130.

Boroughs, D.L., and B. Carpenter, 1991, Helping the Planet and the Economy, *US News and World Report*, 46., 110-11.

Boudoropoulos, I.D., and I.S. Arvanitoyannis, 2000, Potential and Perspectives for Application of Environmental Management System (EMS) and ISO 14000 to Food Industries, *Food Review International*, 16(2), 177-237.

Bowen, H., 1983, Changes in the International Distribution of Resources and Their Impact on US Comparative Advantage, *Review of Economics and Statistics*, 65, 402-414.

Brander, J.A. and B. J. Spencer, 1985, Export Subsidies and International Market Share Rivalry, *Journal of International Economics*, 18, 83-100.

Brown, W., 1990, Trade Deals a Blow to the Environment, *New Scientist*, 10 November.

Brown, L.R., C. Flavin, and S. Postel, 1991, Saving the Planet: How to Shape an Environmentally Sustainable Global Economy, New York: W.W. Norton & Company.

Brown, L.R., C. Flavin, and H. French, 1994, State of the World: 1994, New York: W.W. Norton & Company.

Brown, L.R., C. Flavin, and H. French, 1997, State of the World: 1997, New York: W.W. Norton & Company.

Business Roundtable, 1993, Comparison of Superfund with Programs in Other Countries, Washington, DC: The Business Roundtable.

Carraro, C., and D. Siniscalco, 1993, Strategies for the International Protection of the Environment, *Journal of Public Economics*, 52, 309-328.

Chapman, D., 1991, Environmental Standards and International Trade in Automobiles and Copper: the Case for a Social Tariff, *Natural Resources Journal*, 31(3), 449-61.

Chichilnisky, G., 1994a, Sustainable Development and North-South Trade, Conference Paper, presented at the international conference on "Biological Diversity, Exploring the Complexities" at the University of Arizona, May 1994.

Chichilnisky, G., 1994b, Global Environment and North South Trade, *American Economic Review*, 84, 851-874.

Clark, J. and M.B. Arnold, 1993, The Danish Bottles Case: Commission of the European Communities v. Kingdom of Denmark, *in* the Greening of World Trade, EPA Report no. EPA 100-R-93-002, Washington: EPA.

Cole, M.A., A.J. Rayner, and J.M. Bates, 1997, The Environmental Kuznets Curve: An Empirical Analysis, *Environment and Development Economics*, 2(4), 401-416.

Cole, M.A., Rayner, A.J., and Bates, J.M., 1998, Trade Liberalization and the Environment: the Case of the Uruguay Round, *The World Economy* 21(3), 337-347.

Cole, M.A., 2000, Trade Liberalization, Economic Growth and the Environment, Edward Elgar Publishing.

Conrad, K., 1993, Taxes and Subsidies for Pollution-Intensive Industries as Trade Policy, *Journal of Environmental Economics and Management* 25, 121-135.

Copeland, B.R. 1994, International Trade and the Environment: Policy Reform in a Polluted Small Economy, *Journal of Environmental Economics and Management*, 26, 44-65.

Copeland, B.R., and M. S. Taylor, 1994, North–South Trade and the Environment, *Quarterly Journal of Economics*, 109, 755-787.

Copeland, B.R. and Taylor, M.S., 1995, Trade and the Transboundary Pollution, *American Economic Review*, 85(4), 716-737.

Copeland, B.R. and Taylor, M.S., 1999, Trade, Spatial Separation, and the Environment, *Journal of International Economics* 47, 137-168.

Copeland, B.R. and Taylor, M.S., 2001, International Trade and the Environment: A Framework for Analysis, *NBER Working Papers*, No:8540.

Corden, W., 1974, Trade Policy and Economic Welfare, Clarendon Press, Oxford.

Cropper, M. and Griffith, C. 1994, The Interaction of Population Growth and Environmental Quality, *American Economic Association Papers and Proceedings*, 84(12), 250-254.

D'arge, R., 1974, International Trade, Domestic Income, and Environmental Controls, *in* Managing the Environment, (A. Kneese ed.), New York: Praeger.

Dasgupta, P.S., P.J. Hammond and E.S. Maskin, 1979, The implementation of social choice rules: Some results on incentive compatibility, *Review of Economic Studies*, 46, 185-216.

Dasgupta S., A. Mody, S. Roy, and D. Wheeler, 1995, "Environmental Regulation and Development: A Cross-Country Empirical Analysis" Policy Research Working Paper 1448, Policy Research Department, World Bank Washington, DC.

Dean, J., 1992, Trade and the Environment: A Survey of the Literature, *in* International Trade and the Environment, (P. Low ed.), Washington, DC: World Bank.

Dean, J., 1998, "Testing the Impact of Trade Liberalization on the Environment: Theory and Evidence," mimeo, The Johns Hopkins University.

DeBellevue, E.B., E. Hitzel, E., Cline, K., Benitez, J.A., Ramos-Miranda, J., Segura, O., 1994, The North American Free Trade Agreement: An Ecological-Economic Synthesis for the United States and Mexico, *Ecological Economics* 9(1), 53-71.

Demsetz, H., 1967, Toward a Theory of Property Rights, *American Economic Review*, 57, 347-59.

Dessus, S. and M. Bussolo, 1998, Is There a Trade-off Between Trade Liberalization and Pollution Abatement? A CGE Assessment Applied to Costa Rica, *Journal of Policy Modelling* 20,11-31.

Diwan, I., and Shafik, N., 1992, Investment, Technology and the Global Environment: Towards International Agreement in a World of Disparities, *in* International Trade and the Environment, (P. Low ed.), World Bank, Washington, DC.

Dollar, D., 1992, Outward-Oriented Developing Economies Really do Grow More Rapidly: Evidence from 95 LDCs, 1976-1985, *Economic Development and Cultural Change*, 523-544.

Dorfman, M.H., W.R. Muir, and C.G. Miller, 1992, Environmental Dividends: Cutting More Chemical Wastes, New York: INFORM.

Dornbusch, R., S. Fischer, and P.A. Samuelson, 1977, Comparative Advantage, Trade, and Payments in a Ricardian Model with a Continuum of Goods, *American Economic Review*, 67(5), 823-839.

Duerksen, C., and H.J. Leonard, 1980, Environmental Regulations and the Location of Industries: an International Perspective, *Columbia Journal of World Business*, 15, 52-68.

Edwards, S., 1992, Trade Orientation, Distortions and Growth in Developing Countries, *Journal of Development Economics* 39, 31-57.

Ekins, P., Folke, C., Costanza, R., 1994, Trade, Environment and Development: The Issues in Perspective, *Ecological Economics* 9(1), 1-12.

Ekins, P., 1997, The Kuznets Curve for the Environment and Economic Growth: Examining the Evidence, *Environment and Planning A* 29, 805-830.

Eliste, P., and P.G. Fredriksson, 1998, "Does Open Trade Result in a Race to the Bottom? Cross-Country Evidence", mimeo, The World Bank.

Environmental Protection Agency, 1993, Regulatory Impact Analysis, Stratospheric Protection Division Report, Washington: EPA.

Espinosa, J.A., and K.V. Smith, 1996, Environmental and Trade Policies: Some Methodological Lessons, Resources for the Future Discussion Paper: 96-18.

Esty, D.C., 1994, Greening the GATT: Trade, Environment, and the Future, Washington, DC: Institute for International Economics.

Fankhauser, S., 1995, Valuing Climate Change: The Economics of the Greenhouse, London: Earthscan.

Ferrantino, M.J., 1997, International Trade, Environmental Quality and Public Policy, *The World Economy* 20(1), 43-72.

Ferrantino, M.J., and L.A. Linkins, 1999, The Effect of Global Trade Liberalization on Toxic Emissions in Industry, *Weltwirtschaftliches Archiv* 135(1), 128-155.

Folmer, H., Mouche, P.V., Ragland, S., 1993, Interconnected Games and International Environmental Problems, *Environment and Resource Economics* 3, 313-336.

Francois, J.F., B. McDonald, and H. Nordstrom, 1995, Assessing the Uruguay Round, *in*, The Uruguay Round and the Developing Economies, World Bank Discussion Papers, no. 307, 117-214.

Fredriksson, P.G., 1999, *Trade, Global Policy and the Environment*, The World Bank, Washington, DC.

Fredriksson, P.G., and Gaston, N., 1999, The Importance of Trade for the Ratification of the 1992 Climate Change Convention, In: Fredriksson, P.G., (Ed.), *Trade, Global Policy and the Environment*, The World Bank, Washington, DC.

Friedman, J., D.A. Gerlowski, and J. Silberman, 1992, What Attracts Foreign Multinational Corporations? Evidence from Branch Plant Location in the United States, *Journal of Regional Science*, 32(4), 403-18.

Gardiner, D., 1994, Does Environmental Policy Conflict with Economic Growth, *Resources*, Spring (115), 20-21.

Goodman, S.L., 1998, Is ISO 14001 an Important Element in Business Survival, *The Quality Magazine of Australia*, 6.

Goodstein, E.S., 1995, Economics and the Environment, New Jersey: Prentice-Hall.

Green, J. and J.J. Laffont, 1977, Characterization of Satisfactory Mechanisms for the Revelation of the Preferences for Public Goods, *Econometrica*, 45, 427-438.

Grossman, G.M., and A.B. Krueger, 1991, Environmental Impacts of a North American Free Trade Agreement, National Bureau of Economic Research Working Paper: 3914, November 1991.

Grossman, G.M., and A.B. Krueger, 1993, Environmental Impacts of North American Free Trade Agreement, *in* The US-Mexico Free Trade Agreement, (P. Garber ed.), Cambridge, MA: MIT Press.

Grossman, G.M., and A.B. Krueger, 1995, Economic Growth and the Environment, *Quarterly Journal of Economics*, 110, 353-77

Hansen, B.E. 2000, Sample Splitting and Threshold Estimation, *Econometrica*, 68, 575-603.

Hardin, G., 1968, The Tragedy of Commons, *Science*, 162, 1243-1248.

Harrison, A., 1996, Openness and Growth: A Time-Series, Cross-Country Analysis for Developing Countries, *Journal of Development Economics*, 48, 419-447.

Heal, G., 1994, Formation of International Environmental Agreements, *in* Trade, Innovation, Environment (Ed. C. Carraro), Kluwer Academic Publishers, Dordrecht.

Hester, S., D. Gunasekera, and N. Andrews, 1993, World Market Impacts of EC Restraints on Cropland and Livestock Density, *Agriculture and Resources Quarterly*, 5(1), 71-84.

Hettige, H., R.E.B. Lucas, and D. Wheeler, 1992, The Toxic Intensity of Industrial Production: Global Patterns , Trends and Trade Policy, *American Economic Review, Papers and Proceedings*, 82, 473-481.

Hettige, H., M. Mani, and D. Wheeler, 1995, The Industrial Pollution Projection System World Bank Policy Research Working Paper No. 1431.

Hettige, H., Mani, M., Wheeler D., 1999, Industrial Pollution in Economic Development: Kuznets Revisited, *in* Trade, Global Policy and the Environment, (P. Fredriksson ed.), (World Bank Discussion Papers No. 402), Washington DC, World Bank.

Holtz-Eakin, D.and T.M. Selden, 1995, Stoking the Fires? CO_2 Emissions and Economic Growth, *Journal of Public Economics* 57, 85-101.

Hudson, S., 1992, Trade, Environment and the Pursuit of Sustainable Development, *in* International Trade and the Environment, (P. Low ed.), Washington, DC: World Bank.

Hung, N.M., and Y. Richelle, 1997, Trade Gains, Paretian Transfer, and the Tragedy of the Commons, *The Japanese Economic Review*, 48, 213-225.

Intal, P., E. Medalla, et al., 1994, Trade and Environment Linkages: The Case of the Philippines, Philippine Institute of Development Studies.

International Panel on Climate Change (IPCC), 1990, Scientific Assessment of Climate Change, New York: WMO and UNEP.

Jaffe, A.B., S.R. Peterson, P.R. Portney, and R. N. Stavins, 1995, Environmental Regulation and the Competitiveness of US Manufacturing: What Does Evidence Tell Us, *Journal of Economic Literature*, XXXIII, 132-163.

Jaffe, A., and K. Palmer, 1997, Environmental Regulation and Innovation: A Panel Data Study, *Review of Economics and Statistics*, 79, 610-19

Jaffe, A., R. Newell, and R. Stavins, 2002, Environmental Policy and Technological Change, *Environmental and Resource Economics*, 22, 41-69.

Jha, V., A. Markandya, and R. Vossenaar, 1999, Reconciling Trade and the Environment: Lessons from Case Studies in Developing Countries, Cheltenham:E. Elgar.

Jones, R.W., 1965, The Structure of Simple General Equilibrium Models, *Journal of Political Economy*, 73, 557-72.

Kalt, J.P., 1988, The Impact of Domestic Regulatory Policies on US International Competitiveness, *in* International Competitiveness (M. Spence and H. Hazard eds.), Cambridge: Harper and Row.

Kaufmann, R.K., Davidsdottir B., Garnham S., Pauly P., 1998, The Determinants of Atmospheric SO_2 Concentrations: Reconsidering the Environmental Kuznets Curve, *Ecological Economics*, 25, 209-230.

Kennedy, P.W., 1994, Equilibrium Pollution Taxes in Open Economies with Imperfect Competition, *Journal of Environmental Economics and Management* 27, 49-63.

Khan, M.A., 1996, Free Trade and the Environment, *Journal of International Trade and Economic Development*, 5, 113-36.

Kim, O. and M. Walker, 1984, The Free Rider Problem: Experimental Evidence, Public Choice, 43, 3-24.

Knodgen, G., 1979, Environment and Industrial Siting: Results of an Empirical Survey of Investment by West German Industry in Developing Countries, *Zeitschrift fur Umweltpolitik*, 2.

Kraus, C., 2000, Import Tariffs as Environmental Policy Instruments, Dordrecht:Kluwer Academic Publishers.

Krutilla, K., 1991, Environmental Regulation in an Open Economy, *Journal of Environmental Economics and Management*, 20, 127-142.

Lanjouw, J.O., and A. Mody, 1996, Innovation and the International Diffusion of Environmentally Responsive Technology, *Research Policy*, 25, 549-571.

Lee, H., and D. Roland-Holst, 1994, International Trade and Transfer of Environmental Costs and Benefits, *in* Applied Trade Policy Modeling, (J. Francois and K. Reinert eds.) Cambridge: Cambridge University Press.

Lee, H. and Roland-Holst, D., 1997, The Environment and Welfare Implications of Trade and Tax Policy, *Journal of Development Economics* 52, 65-82.

Lee, J.R., 1996, Basic Attributes of Trade and Environment: What do the Numbers Tell Us? *Ecological Economics*, 19, 19-33.

Levinson, A., 1992, Environmental Regulations and Manufacturers' Location Choices: Evidence from the Census of Manufactures, New York: Columbia University.

Levinson, A., 1996, Environmental Regulations and Industry Locations: International and Domestic Evidence, *in* Fair Trade and Harmonization (J. Bhagwati and R. Hudec eds.) Cambridge: The MIT Press.

Lopez, R., 1994, The Environment as a Factor of Production: The Effects of Economic Growth and Trade Liberalization, *Journal of Environmental Economics and Management*, 27, 163-184.

Lopez, R., 1997, Environmental Externalities in Traditional Agriculture and the Impact of Trade Liberalization: The Case of Ghana, *Journal of Development Economics*, 53, 17-39.

Low P., 1991, Trade Measure and Environmental Quality: The Implications for Mexico's Exports. Washington: World Bank, mimeo.

Low, P., 1992, International Trade and the Environment, Washington, DC: World Bank.

Low, P., and R. Safadi, 1992, Trade Policy and Pollution, *in* International Trade and the Environment, (P. Low ed.), Washington, DC: World Bank.

Low, P., and A. Yeats, 1992, Do "Dirty" Industries Migrate, *in* International Trade and the Environment, (P. Low, ed.), Washington, DC: World Bank.

Lucas, R.E.B., Wheeler, D., Hettige, H., 1992, Economic Development, Environmental Regulation, and the International Migration of Toxic Industrial Pollution: 1960-1988. *in* International Trade and the Environment, (P. Low, ed.), World Bank, Washington, DC.

Magee, S.,and W.F. Ford, 1972, Environmental Pollution, the Terms of Trade, and the Balance of Payments, *Kyklos*, 25, 101-18.

Mäler, K.G., 1990, International Environmental Problems, *Oxford Review of Economic Policy*, 6(1), 80-108.

Mani, M., and D. Wheeler, 1997, "In Search of Pollution Havens? Dirty Industry Migration in the World Economy," World Bank Working Paper, No:16.

Markussen, J.R., 1975, International Externalities and Optimal Tax Structure, *Journal of International Economics*, 5, 15-29.

McGuire, M.C., 1982, Regulation, Factor Rewards, and International Trade, *Journal of Public Economics*, 17 (3), 335-54.

Merrifield, J.D., 1988, The Impact of Selected Abatement Strategies on Transnational Pollution, the Terms of Trade, and Factor Rewards: A General Equilibrium Approach, *Journal of Environmental Economics and Management*, 15, 259-84.

Montgomery, W.D., 1972, Markets in Licenses and Efficient Pollution Control, *Journal of Economic Theory*, 5(3), 395-418.

Mytelka, L.K., 1999, Competition, Innovation and Competitiveness in Developing Countries, OECD Press: Paris.

Nordstrom, H., and S. Vaughan, 1999, Trade and Environment, World Trade Organization.

Oates, W. E., K. P. Palmer and P. R. Portney, 1994, Environmental Regulation and International Competitiveness: Thinking About the Porter Hypothesis, Resources for the Future Discussion Paper, 94-02.

OECD, 1978, Macroeconomic Evaluation of Environmental Programmes, OECD Press: Paris.

OECD, 1991, The State of the Environment, OECD Press: Paris.

Palmer, K. P., W. E. Oates, and P. R. Portney, 1995, Tightening Environmental Standards: The Benefit-Cost or No-Cost Paradigm, *Journal of Economic Perspectives*, 9(4), 119-132.

Panayotou, T., 1993a, Green Markets: The Economics of Sustainable Development, San Francisco: ICS Press for the International Center for Economic Growth and the Harvard Institute for International Development.

Panayotou T., 1993b, "Empirical Tests and Policy Analysis of Environmental Degradation at Different Stages of Economic Development", Working Paper WP238, Technology and Employment Programme, International Labour Office, Geneva.

Panayotou T., 1997, Demystifying The Environmental Kuznets Curve: Turning A Black Box Into A Policy Tool , *Environment and Development Economics*, 2(4), 465-484.

Parkinson, G., 1990, Reducing Wastes Can be Cost-Effective, *Chemical Engineering*, 97, 30.

Pasurka, C.A., 1985, Environmental Control Costs and U.S. Effective Rates of Protection, *Public Finance Quarterly*, 13(2), 161-182.

Pearce, D.W., 1993, Blueprint Three: Measuring Sustainable Development, London: Earthscan.

Pearson, C., 1987, Multinational Corporations, Environment and the Third World, Durham: Duke Univ. Press.

Pethig, R., 1976, Pollution, Welfare and Environmental Policy in the Theory of Comparative Advantage, *Journal of Environmental Economics and Management*, 2, 160-69.

Petrakis, E., and A. Xepapadeas, 1996, Environmental Consciousness and Moral Hazard in International Agreements to Protect the Environment, *Journal of Public Economics*, 60, 95-110.

Piggott, J., J. Whalley, and R. Wigle, 1992, International Linkages and Carbon Reduction Initiatives, *in* The Greening of World Trade Issues, (K. Anderson and R. Blackhurst eds.), Ann Arbor: University of Michigan Press.

Porter, M. A., 1991, America's Green Strategy, *Scientific American*, 168, 264.

Porter, M. A., and C. van der Linde, 1995, Towards a New Conception of the Environment-Competitiveness Relationship, *Journal of Economics Perspectives*, 9(4), 97-118.

Ratnayake, R., 1998, Do Stringent Environmental Regulations Reduce the International Competitiveness? Evidence from an Inter-Industry Analysis, *International Journal of Economics of Business*, 5, 77-96.

Rauscher, M., 1994, On Ecological Dumping, *Oxford Economic Papers*, 46, 822-40.

Rauscher, M., 1997, International Trade, Factor Movements and the Environment, Oxford University Press: Oxford.

Richardson, J.D., and J.H. Mutti, 1976, Industrial Development through Environmental Controls: The International Competitive Aspect, *in* Studies in International Environmental Economics, (I. Walter ed.), New York: Wiley.

Richardson, J.D., and J.H. Mutti, 1977, International Competitive Displacement from Environmental Control, *Journal of Environmental Economics and Management*, 4, 135-52.

Ritchie, M., 1990, GATT, Agriculture and the Environment: the US Double Zero Plan, *The Ecologist*, 20(6), 214-20.

Robinson, H.D., 1988, International Pollution Abatement: the Impact on the Balance of Trade, *Canadian Journal of Economics*, 21, 187-99.

Rodrik, D., 1994, Getting Interventions Right: How South Korea and Taiwan Grew Rich, *NBER Working Papers*, No: 4964.

Rothman, D.S., 1998, Environmental Kuznets Curves-Real Progress or Passing the Buck? A Case for Consumption-Based Approaches, *Ecological Economics* 25, 177-194.

Runge, F.C., F. Ortalo-Magne, and P.V. Kamp, 1994, Freer Trade, Protected Environment, Council on Foreign Relations Press: New York.

Sachs, J.D., and Warner, A., 1995, "Economic Reform and the Process of Global Integration", *Brookings Papers on Economic Activity*, 1, 1-119.

Samuelson, P., 1962, The Gains from International Trade Once Again, *Economic Journal*, 72, 820-829.

Sartzetakis, E. S., and C. Constantatos, 1995, Environmental Regulation and International Trade, *Journal of Regulatory Economics*, 8(1), 61-72.

Schmalensee, R., T.M. Stoker, and R. Judson, 1998, World Carbon Dioxide Emissions: 1950-2050, *Review of Economics and Statistics*, 80(1), 15-27.

Selden, T.M. and D. Song, 1994, Environmental Quality and Development: Is There a Kuznets Curve for Air Pollution Emissions?, *Journal of Environmental Economics and Management*, 27, 147-162.

Shafik, N., and S. Bandyopadhyay, 1992, "Economic Growth and Environmental Quality: Time Series and Cross-Country Evidence", World Bank Policy Research Working Paper, WPS 904, Washington DC: World Bank.

Shafik N., 1994, Economic Development and Environmental Quality: An Econometric Analysis, *Oxford Economic Papers*, 46, 757-773.

Shrybman, S., 1990, International Trade and the Environment: An Environmental Assessment of the GATT, *The Ecologist*, 20(1), 30-31.

Siebert, H., 1977, Environmental Quality and Gains From Trade, *Kyklos*, 30 (4), 657-73.

Siebert, H., 1981, Economics of the Environment, Cambridge: Lexington Books.

Simpson, R. D., and R. L. Bradford, 1996, Taxing Variable Cost: Environmental Regulation as Industrial Policy, *Journal of Environmental Economics and Management*, 30, 282-300.

Smith, K., and A. Espinosa, 1996, Environmental and Trade Policies: Some Methodological Lessons, Resources For the Future Discussion Paper, 96-18.

Smith, V., 1979, An Experimental Comparison of Three Public Good Decision Mechanisms, *Scandinavian Journal of Economics*, 81(2), 198-215.

Strutt, A., and Anderson, K., 1999, Will Trade Liberalization Harm the Environment? The Case of Indonesia to 2020, *in* Trade, Global Policy and the Environment, (P. Fredriksson ed.), (World Bank Discussion Papers No. 402), The World Bank, Washington, DC.

Summers, R. and A. Heston, 1991, The Penn World Table: An Expanded Set of International Comparisons, 1950-1988, *Quarterly Journal of Economics*, 106(2), 327-368.

Suri, V., Chapman, D., 1998, Economic Growth, Trade and Energy: Implications for the Environmental Kuznets Curve, *Ecological Economics*, 25, 195-208.

Sweeney, J., 1973, An Experimental Investigation of Free Rider Problem, *Social Science Research*, 2, 227-292.

Tansey, G. and T. Worsley, 1995, The Food System: A Guide, London:Earthscan Publ. Ltd.

Taskin, F. and O. Zaim, 2001, The Role of International Trade on Environmental Efficiency: A DEA Approach, *Economic Modelling*, 18, 1-17.

Tobey, J., 1990, The Effects of Domestic Environmental Policies on Patterns of World Trade: an Empirical Test, *Kyklos*, 43, 191-209.

Tinbergen, J., 1952, On the Theory of Economic Policy, Amsterdam: North Holland Press.

Torras M. and J. K. Boyce, 1996, Income, Inequality, and Pollution: A Reassessment of the Environmental Kuznets Curve, *Ecological Economics*, special issue on EKC.

Tsai, P.L., 1999, Is Trade Liberalization Harmful to the Environment? An Alternative View, *Journal of Economic Studies* 26(3), 201-108.

Ugelow, J., 1982, A Survey of Recent Studies on Costs of Pollution Control and the Effects on Trade, in Environment and Trade, (S. Rubin ed.), New Jersey: Allanheld, Osmun and Co.

Ulph, A., 1990, The Choice of Environmental Policy Instruments and Strategic International Trade, University of Southampton Discussion Paper in Economics and Econometrics, 9012.

Ulph, A., 1994, Environmental Policy and International Trade–A Survey of Recent Economic Analysis, University of Southampton Discussion Paper in Economics and Econometrics, 9423.

Ulph, A., 1996, Environmental Policy and International Trade when Governments and Producers Act Strategically, *Journal of Environmental Economics and Management* 30, 265-281.

UN Food and Agricultural Organization (UNFAO), 1993, World Review of High Seas and Highly Migratory Fish Species and Straddling Stocks, UNFAO Fisheries Circular No. 858, Rome.

US Department of Commerce, 1975, The Effects of Pollution Abatement on International Trade-III.

Van Beers, C. and J.C.J.M. van den Bergh, 1996, An overview of Methodological Approaches in the Analysis of Trade and Environment, *Journal of World Trade* , 30(1), 143-167.

Van Beers, C. and J.C.J.M. van den Bergh, 1997, An Empirical Multi-Country Analysis of the Impact of Environmental Policy on Foreign Trade Flows, *Kyklos*, 50(1), 29-46.

Walter, I., 1973, The Pollution Content of American Trade, *Western Economic Journal*, 11, 61-70.

Walter, I., and J. Ugelow, 1979, Environmental Policies in Developing Countries, *Ambio*, 8, 102-09.

Walter, I., 1982, Environmentally Induced Industrial Relocation to Developing Countries, *in* Environment and Trade, (S. Rubin ed.), New Jersey: Allanheld, Osmun, and Co.

World Bank, 1992, World Tables 1992, Baltimore: Johns Hopkins University Press.

World Resources Institute, 1992, World Resources 1992-93, New York: Oxford University Press.

Xepapadeas, A., and A. De Zeeuw, 1999, Environmental Policy and Competitiveness: The Porter Hypothesis and the Composition of Capital, *Journal of Environmental Economics and Management*, 37, 165-82.

Xing, Y., and C. Kolstad, 2002, Do Lax Environmental Regulations Attract Foreign Investment?" *Environmental and Resource Economics*, 21, 1-22.

Xu, X., 1999, Do Stringent Environmental Regulations Reduce the International Competitiveness of Environmentally Sensitive Goods? A Global Perspective, *World Development*, 27, 1215-26.

Yezer, A., and A. Philipson, 1974, Influence of Environmental Considerations on Agriculture and Industrial Decisions to Locate Outside of the Continental Us., Public Interest Economics Center.

Yohe, G. W., 1979, The Backward Incidence of Pollution Control-Some Comparative Statics in General Equilibrium, *Journal of Environmental Economics and Management*, 6(3), 187-98.

Yu, E.S.H., and C.A. Ingene, 1982, The Backward Incidence of Pollution Control in a Rigid-Wage Economy, *Journal of Environmental Economics and Management*, 9, 304-310.

Index

FSC
www.fsc.org
MIX
Papier | Fördert gute Waldnutzung
FSC® C083411

Kolibri 360 GmbH
Ferdinand-Jühlke-Straße 7
99095 Erfurt, Deutschland
produktsicherheit@kolibri360.de